"Economides's *Wild Anthropocene* engages deep time with brilliance, nuance, and attention to detail. Her lucid analyses of both environmental poetry and fiction persuasively demonstrate the power of literature in rethinking the human and positioning different human communities in relation to geological time, biodiversity loss, and climate change. A fascinating read for anyone interested in the role of literature in its engagement with the global eco-crises of the Anthropocene."

Ursula K. Heise, *Distinguished Professor, Department of English/Institute of the Environment and Sustainability, University of California, Los Angeles, USA*

"The Anthropocene is a dangerous concept: in shattering traditional conceptions of the human, it also puts familiar understandings of political agency at stake. In her wide-ranging readings of texts by authors as different as Margaret Atwood, Dionne Brand, Robinson Jeffers, and Jeff Vandermeer, Louise Economides shows how literature's engagement with deep time can energize a politics of social and ecological justice for this moment of peril. Deeply insightful and elegantly written, her book is an invaluable contribution to ongoing debates about the Anthropocene in the environmental humanities."

Hannes Bergthaller, *National Taiwan Normal University*

"This incisive ecocritical study explores deep time imaginaries in recent North American poetry and fiction. Economides shows how literary refractions of evolutionary and geological timescales can illuminate the stakes of political struggle and the horizons of justice in the Anthropocene."

Tobias Menely, *University of California, Davis, USA*

"Countering 'ecomodernist optimism's' obliviousness to its racist and sexist histories and to the ongoing experiences of global environmental injustice, Louise Economides's *Wild Anthropocene: Literature and Multispecies Justice in Deep Time* offers readers a powerful alternative with her embrace of the strange, perhaps even partially artificial, wildness rampant in the Anthropocene. She avoids both the flat ontology attributing full agency to all forces of earthly ecologies and the corporate-friendly belief that Anthropocene engineers can control the massively shifting ecologies of Earth. Instead, *Wild Anthropocene* traces literature's contributions to shock, critique, and trace emergences of promising human and non-human networks."

Heather I. Sullivan, *Professor of German and Comparative Literature, Trinity University, USA*

Wild Anthropocene

Wild Anthropocene examines four key areas—the politics of deep time, neoliberalism's socio-ecological impacts, global population growth and inter-species entanglement—to demonstrate how literature illuminates progressive solutions to Anthropocene challenges. The book argues that technological mitigation of contemporary environmental crises must be complemented by a politics committed to multispecies justice. Central to this new politics is the project of reimagining our relationship with time as something other than its status within capitalist praxis.

The book brings together poetry and fiction written by a diverse range of writers to demonstrate how contemporary literature addresses important connections between social oppression and environmental issues. It also critiques techno-managerial visions of the future that celebrate humanity's ever-growing "control" over ecosystems by examining multiple sources of wildness (temporal, environmental and technological) that expose the problematic ideology underwriting such aspirations. Readers will be introduced to a way of understanding the Anthropocene that, while being informed by recent discoveries in Earth science and evolutionary biology, also makes a strong case for humanities-based understanding of environmental politics.

This interdisciplinary text will be a useful addition to theoretical discussions on the Anthropocene for scholars, researchers and students in the environmental humanities, literary studies, ecocriticism and environmental philosophy.

Louise Economides is a Professor of English and Director of Literature and the Environment at the University of Montana, USA.

Routledge Environmental Humanities

Series editors: Scott Slovic (University of Idaho, USA), **Joni Adamson** (Arizona State University, USA) and **Yuki Masami** (Aoyama Gakuin University, Japan)

The *Routledge Environmental Humanities* series is an original and inspiring venture recognising that today's world agricultural and water crises, ocean pollution and resource depletion, global warming from greenhouse gases, urban sprawl, over-population, food insecurity and environmental justice are all *crises of culture.*

The reality of understanding and finding adaptive solutions to our present and future environmental challenges has shifted the epicenter of environmental studies away from an exclusively scientific and technological framework to one that depends on the human-focused disciplines and ideas of the humanities and allied social sciences.

We thus welcome book proposals from all humanities and social sciences disciplines for an inclusive and interdisciplinary series. We favour manuscripts aimed at an international readership and written in a lively and accessible style. The readership comprises scholars and students from the humanities and social sciences and thoughtful readers concerned about the human dimensions of environmental change.

For more information about this series, please visit: www.routledge.com/Routledge-Environmental-Humanities/book-series/REH

Wild Anthropocene

Literature and Multispecies Justice in Deep Time

Louise Economides

Routledge
Taylor & Francis Group
LONDON AND NEW YORK

earthscan
from Routledge

First published 2025
by Routledge
4 Park Square, Milton Park, Abingdon, Oxon OX14 4RN

and by Routledge
605 Third Avenue, New York, NY 10158

Routledge is an imprint of the Taylor & Francis Group, an informa business

British Library Cataloguing-in-Publication Data
A catalogue record for this book is available from the British Library

Library of Congress Cataloging-in-Publication Data
Names: Economides, Louise - author.
Title: Wild anthropocene : literature and multispecies justice in deep time / Louise Economides.
Description: New York : Routledge, 2025. | Series: Routledge environmental humanities | Includes bibliographical references and index.
Identifiers: LCCN 2024041305 (print) | LCCN 2024041306 (ebook) | ISBN 9781032153902 (hardback) | ISBN 9781032153926 (paperback) | ISBN 9781003243946 (ebook)
Subjects: LCSH: Time in literature. | Environmental justice in literature. | Ecocriticism. |
LCGFT: Literary criticism. | Essays.
Classification: LCC PN56.T5 E26 2025 (print) | LCC PN56.T5 (ebook) | DDC 820.93384–dc23/eng/20241119
LC record available at https://lccn.loc.gov/2024041305
LC ebook record available at https://lccn.loc.gov/2024041306

ISBN: 978-1-032-15390-2 (hbk)
ISBN: 978-1-032-15392-6 (pbk)
ISBN: 978-1-003-24394-6 (ebk)

DOI: 10.4324/9781003243946

Typeset in Galliard
by Taylor & Francis Books

For my parents and future generations

Contents

Acknowledgements

Being the product of turbulent times in my personal life and in the world at large, many thanks are due to those who lent a hand in this book's production. At the University of Montana, I appreciate input I received about the project from faculty and students in the Environmental Humanities Working Group. Likewise, I am thankful to departmental and college administrators who supported a flexible work schedule that allowed me to stay on schedule writing the book, despite challenges that arose due to the Covid pandemic. Warm thanks also go out to Victoria Boyer and other staff (and faculty) at the Flathead Lake Biological Station for assisting with annual visits to this facility, where I went to write while immersed in the tranquil beauty of Yellow Bay. Special thanks to Jim and Paula Karman for their invaluable scholarly expertise regarding Robinson Jeffers and for their advice about the book's first chapter. Likewise, I am indebted to Aaron Yoshinobu for his insights about deep time, the Anthropocene and Jeffers's poetry from the perspective of a professional geologist. I'm also very grateful to my editor at Routledge, Matthew Shobbrook, for his patience throughout the production process and for his sensitivity regarding health challenges I navigated along the way. And, as always, I am forever indebted to my partner Rob Browning for inspiring me to explore literature through a "big picture" lens, for his equally impactful attentiveness to details in my book and for his never-flagging encouragement that I complete this work.

I'd also like to thank Lindsay Jeffers and Stanford University Press for granting permission to publish Jeffers's poems. "Carmel Point," Copyright © 1954 by Robinson Jeffers and renewed 1982 by Donnan Call Jeffers & Garth Jeffers; from THE SELECTED POETRY OF ROBINSON JEFFERS by Robinson Jeffers. Used by permission of Random House, an imprint and division of Penguin Random House LLC. All rights reserved. "Loss Creek," "Devonian," and "Gneiss" from STRIKE/SLIP by Don McKay, Copyright © 2006 Don McKay. Reprinted by permission of McClelland & Stewart, a division of Penguin Random House Canada Limited. All rights reserved. Thanks to Gaspereau Press for permission to publish lines from "Between Rock and Stone, A Geopoetic Alphabet" from DEACTIVATED WEST 100, Copyright © 2005 Don McKay. "IV xiii," "XIII," and "XV" from LAND TO LIGHT ON by Dionne Brand, Copyright@ 1997 Dionne Brand. Reprinted by permission of McClelland & Stewart, a division of Penguin Random House Canada Limited. All rights reserved. "Ossuary I," "Ossuary XV," "Ossuary XII," and "Ossuary XIV," from OSSUARIES by Dionne Brand, Copyright © 2010 Dionne Brand. Reprinted by permission of McClelland & Stewart, a division of Penguin Random House Canada Limited. All rights reserved. Excerpts from LAND TO LIGHT ON by Dionne Brand, Copyright © 1997 by Dionne Brand, INVENTORY by Dionne Brand, Copyright @ 2006 by Dionne Brand, and OSSUARIES by Dionne Brand, Copyright © 2010 by Dionne Brand, used by permission of The Wylie Agency LLC.

Introduction: Towards a Wild Anthropocene

Imagining Multispecies Justice in Deep Time

Although the title of Werner Herzog's *Cave of Forgotten Dreams* announces it will explore the visions of paleolithic humans who left glorious paintings on walls of the Chauvet Cave over 30,000 years ago, the documentary undertakes something more than this: nothing less than a deep time journey that encompasses our species' prehistoric past, more recent Holocene history and emerging Anthropocene future. This journey begins with an opening shot of a vineyard which suggests the Holocene period, a time of relatively stable climate in which the agricultural revolution and global civilizations arose. Suggestively, from this ground-level perspective we could be in a vineyard almost anywhere; it is only when the camera rises into the air and we see the Pont-d'Arc looming on the horizon that we realize we are in southern France. The camera takes us through this limestone landmark as through a portal to another world: we are headed to an ice-age Europe covered by steppes and populated by megafauna—rhinos, bison, aurochs, wolves, lions and cave bears—as well as *Homo sapiens* and Neanderthals adapted to living in a climate that was generally colder than today's and that fluctuated between cooling and warming periods known as Dansgaard-Oeschger cycles. In caves near the stone arch, paleolithic peoples painted megafauna in motion such as bison with eight legs that bear an uncanny resemblance to animated film stills, figures designed to come to life by flickering firelight. Lulled into a kind of trance by Ernst Reijseger's choral soundtrack, Herzog's signature Teutonic musings and images that are so astonishingly vibrant that their authenticity had to be verified by the presence of calcites/concretions on top that could only have formed over millennia, viewers are transported to a forgotten animistic past when, we are told, there was a greater sense of permeability between humans, other living beings and the spiritual world.

This deep time reverie is abruptly punctured at the end of the documentary when a drone camera is shown returning to Herzog's film crew on a sandy beach near the Pont-d'Arc, immediately followed by a shot of the nearby Rhone River nuclear power plant. Here, we are taken on a final surreal journey inside the walls of huge greenhouses heated by warm water used to cool the plant. Within these structures, an artificial tropical biosphere flourishes on top of land that was once covered by glaciers. We are told that within the "new climate" of this biosphere, crocodiles—even mutant albino varieties—are flourishing. The evocation of nuclear power, artificial ecosystems and anthropogenic climate (synecdoche for a

DOI: 10.4324/9781003243946-1

greenhouse planet) all signal that we have now entered a new era: the Anthropocene. Herzog's voiceover suggests that the white crocs thriving in this artificial environment may be doppelgangers of Anthropocene-era humanity, asking: "Are we today, possibly, the crocodiles who look back into an abyss of time when we see the paintings of Chauvet Cave?" Perhaps, instead of identifying with these traces of long-lost animism, contemporary humans might find them as alien and incomprehensible as paleolithic peoples would find our ways of thinking about the natural world. It is tempting to read the documentary's temporal arc as one that traces our species' fall from a state of early spiritual reverence for the more-than-human world to today's alienating technosphere in which humans now presume to determine the scope and future of "nature." Yet, a study of *Homo sapiens'* deep history and present ecological conundrum reveals the inadequacies of such a narrative. For example, prehistoric peoples similar to those who painted the Chauvet Cave may have been responsible for waves of megafauna extinction, first 40,000 years ago in Australia and then 15,000 years ago in North and South America.[1] Do the beautiful images of such animals on Chauvet's walls reflect paleolithic reverence or an early quest to attain a kind of power over such creatures via totemic symbols? Likewise, our present Anthropocene relationship with the natural world is (arguably) less reflective of Promethean mastery than it is of Epimethean buffoonery as we try to devise ever-more-powerful technologies to contain multiple, cascading ecological crises of our own making—less a reflection of Bacon's laboratories in Salomon's House than of Mickey Mouse's struggles to keep a flood at bay in *Fantasia*'s "The Sorcerer's Apprentice." In an era of anthropogenic global warming, we are returning to a pre-Holocene state of climactic instability and increasingly chaotic positive feedback loops in the Earth's ecological systems. Perhaps Herzog's focus on primordial reptiles thriving in the Rhone plant's biosphere is a way of suggesting that an element of recalcitrant Anthropocene *wildness* will persist and may even expand, however synthetic this new "human" epoch may become.

Although there have been many objections to using the term "Anthropocene" as a label for the crossroads at which we find ourselves today,[2] one of the more deeply meaningful debates concerns whether, in order to protect future ecosystems, we must accept a new paradigm that moves from passive conservation to active construction: from saving what remains of already existing "nature" to redesigning it to better meet our species' needs and others we deem worthy of protection. To regard techno-managerial accounts of the Anthropocene with the degree of hermeneutic suspicion that I have registered in my discussion of *Cave of Forgotten Dreams* is to challenge ecomodernist faith that *Homo sapiens* not only can but *should* deliberately engineer the ecosystems of tomorrow. Faced with what French historians Christophe Bonneuil and Jean-Baptiste Fressoz describe as the "shock" of acknowledging that, on a species level, humankind's collective agency upon the Earth is now analogous to a geological-scale force, ecomodernist thinkers such as Erle Ellis, Emma Marris, Diane Ackerman, Peter Kareiva, Joseph Mascaro, Stewart Brand and others seek to transmute leaden environmental dread into golden technological optimism. Such theorists reason that if it is the case we moderns have inadvertently had an enormous global impact on the planet then we now have a responsibility to guide these impacts more *deliberately*, embracing our role as geological agents by creating designer ecosystems, engaging in de-extinction and rewilding schemes, and developing geoengineering, bioengineering and other emergent technologies. The story goes that, with the right "can do" attitude, we can engineer our way out of large-scale and complex ecological problems such as climate change, ocean acidification, mass extinction, desertification and increased risk of pandemics. Science writer Emma Marris succinctly sums up the new ecomodernist mantra:

> We've forever altered the Earth, and so now we cannot abandon it to a random fate. It is our duty to manage it. Luckily, it can be a pleasant, even joyful task if we embrace it in the right spirit. Let the rambunctious gardening begin.
>
> (*Rambunctious Garden* 171)

It is important to note that such a techno-managerial ethos was, from the outset of Anthropocene theory, a logical conclusion that could be drawn about our collective responsibility in the face of novel environmental challenges. Though less exuberant than Marris's *Rambunctious Garden*, meteorologist and atmospheric chemist Paul Crutzen's ground-breaking article "Geology of Mankind" makes a similar point:

> Unless there is a global catastrophe … mankind will remain a major environmental force for many millennia. A daunting task lies ahead for scientists and engineers to guide society towards environmentally sustainable management during the era of the Anthropocene. This will require appropriate human behaviour at all scales, and may well involve internationally accepted, large-scale geo-engineering projects, for instance, to 'optimize' climate.
>
> (23)

Undeniably, there are certain advantages to ecomodernism's blueprints for the future. Proponents of this ethos are right to envision a sweeping global shift away from fossil fuels to more sustainable energy sources (including solar power), and proposals to centralize future urban centers by moving away from sprawling suburbs in favor of "intensive" housing and agriculture to make more room for pastoral and wild spaces make immanent sense.[3] Yet some "large-scale geoengineering projects" designed to "'optimize'" our warming climate—such as proposals to shoot tons of aerosols into the Earth's stratosphere in order to dim sunlight to produce a cooling effect that mimics the aftermath of volcanic eruptions (aka "stratospheric aerosol injection")—seem at best reckless in their underestimation of unforeseen side-effects from tinkering with complex, dynamic systems and, worse, as proposals that might actually discourage a reduction of fossil fuel emissions if such "temporary" measures became permanent policy for reasons of financial expediency.[4] Do you enjoy the sight of blue sky? Imagine a world where this was *never* visible, and you will begin to grasp the implications of the brave new world we're heading towards.[5]

Defenders of what I'm calling the "techno-managerial" Anthropocene often represent themselves as realists who accept (and even celebrate) the responsibility of increasing environmental intervention over the objections of outmoded "romantic" ecologists. They often frame the grounds of debates about what conservation should look like in the future as a battle between their own enlightened pragmatism versus older moral fundamentalism. Instead of fetishizing wilderness, thinkers like Marris encourage us to see "nature almost everywhere" and to "temper our romantic notion of untrammeled wilderness and find room next to it for the more nuanced notion of a global, half-wild rambunctious garden, tended by us" (2). At first glance, such an ethos might appear to be a progressive alternative to romantic conservation's potentially paralyzing eco-melancholic nostalgia. Marris cites Bill McKibben's *The End of Nature* as a prime example of the latter, in its mourning for the loss of "untouched" ecosystems in the climate-change era, and its gloomy conclusion that "there is no future in loving [this kind of] nature" (McKibben 180). In "Humility in the Anthropocene," Marris critiques such "old-school" conservationism as seeking to "make [the] green future *all about us*," about protecting only those more-than-human phenomena we associate with ecological "virtue" (47, emphasis in original). Instead of mourning for the loss of nature's "lost dignity and wildness" as thinkers like McKibben do, Marris advocates for a new do-whatever-works "biocentrism." Discussing species and habitat loss in the Anthropocene, she observes: "Lots of endangered plants are happy as pigs in cities. So yes, we can and should protect their native habitat. But also—why not?—plant them in T.J. Maxx parking lots" (47–8).

Yet despite Marris's well-intended critique of "game-over" ecological mourning, her corporate-friendly examples of "rambunctious gardening" belie what is lacking in techno-managerial ecology as a viable praxis. Ecomodernist optimism often relies upon a marked blindness to environmental and social injustice, to the ways that power inequities have shaped the history of technological development and ecological policy under capitalist modernity. By framing the grounds of the debate about possible Anthropocene futures as a "moral" struggle between "hands off" wilderness reactionaries and progressive technocratic gardeners, Marris and other ecomodernists downplay or outright ignore daunting political and economic barriers to the emergence of truly responsible praxis; typically such challenges and the difficult choices they require disappear in a fog of generalization. Who is the "we" that Marris imagines as being always-already committed to protecting endangered species, and who might also be decorating T.J. Maxx parking lots with such life (would such individuals be from the same socio-economic class and would they share the same values)? And what if TJX Companies, inc., who owns T.J. Maxx, decided it didn't want to spend money decorating their parking lots with endangered species, or on maintaining them? What if, instead, they wanted to open many new stores (all with parking lots!) built on areas currently designated as endangered habitats that sustain not only the species "we" have selected for survival but myriad other creatures? What utility would there be for a company like T.J. Maxx, with a history of exploitative labor practices and questionable environmental policies, to preserve endangered species beyond shallow greenwashing campaigns? This is not to say that corporations cannot strive (sometimes successfully) to improve their environmental sustainability; rather, I am asserting that corporate labor and environmental policies must be part of the conversation, in contrast to technocratic silence on such matters.

Other ecomodernist thinkers such as Erle Ellis (an environmental scientist) reflect similar political naïveté while making more overtly hubristic claims than Marris about our species' future role as planetary engineers. In "Too Big for Nature," Ellis proclaims

> To love nature in the Anthropocene it will be necessary to love an artificial nature, to cherish artificial wilderness along with artificial evolution. In the Anthropocene, humans do not disturb nature. We re-shape it. The age-old view of humans as destroyers of nature no longer holds. The Anthropocene demands that we view humanity in the act of creating and sustaining a new nature—and one that will endure in geologic time. Only to the extent that humanity is able to engineer, design, and conserve nature more actively than ever before will humanity or nonhuman nature thrive in the Anthropocene. The question is now not how nature can continue to sustain humanity, but how humanity can continue to sustain nature.
>
> (29)

Again, there is no specificity in Ellis's ecomodernist manifesto about which "humans" are going to be empowered to "re-shape" a "new nature" in the future, or how these engineers will avoid past and present mismanagement of natural resources. Given accelerating extinction rates in non-human species and the cascading effects of climate change—incredibly destructive superstorms, massive fires, flooding events, ocean acidification—it seems premature to announce that "the age-old view of humans as destroyers of nature no longer holds." Many humanities scholars have rightly deconstructed problematic universalism at work in the technocratic Anthropocene, identifying ways in which thinkers like Ellis not only erase the history and ongoing impacts of fossil capitalism and extractive colonialism's "engineering" of the Earth, but also—because of such lacunas—risk reinforcing ideological oppression associated with these energy regimes, including classism, racism, sexism, erotophobia, ableism and species-based discrimination.

Yet despite ecomodernism's shortcomings there is one area where a quasi-managerial mindset may not only be helpful but even necessary: namely, adjustments to what politics might mean within the context of deep time. It is significant that Ellis calls for greater deliberateness not only in the project of "creating and sustaining a new nature" but also of insuring that this new reality will be "one that will endure in geologic time." A curious addition to an already ambitious list, which one might be tempted to see as another example of bombastic overconfidence. After all, as the documentary *Into Eternity* reveals via exploring Onkalo, a massive Finnish nuclear waste repository, no human structure in history has ever endured for one-tenth of the 100,000 years required to safely house this waste—a temporal blink of an eye in terms of geological timescales, which are typically on the order of millions or billions of years. So, what could Ellis possibly mean when he claims that our species, a relative newcomer on the planet in evolutionary terms, now has the power to "shape" the future in such a profound way?

It is critically important to acknowledge that there is no way to adequately answer this question if we attempt to do so from the perspective of separate disciplinary silos: the gap between those working in the sciences versus the humanities. To begin to answer this question requires not just a working knowledge of geology, evolutionary biology, climate science and recent technological advances, but also of political, economic, religious and psychological sources of *resistance* to thinking about human agency as something that can be meaningfully connected to deep time. What does "responsibility" look like from the vantage point of geological-scale force? And what roles might literature play in illuminating this responsibility? This book will investigate these questions in the hope that the arc of human history does not begin and end in darkness. Despite their many differences, there is a strange symmetry to the stories told in *Cave of Forgotten Dreams* and *Into Eternity* insofar as both reflect the power and promise of art and technology but also a capacity for violence against nature, autocratic power and human exceptionalism. Although many animal bones were found in the Chauvet Cave, no human remains have ever been located in them, suggesting that our ancestors never lived in these structures but cordoned them off for ceremonies conducted by a select few—perhaps shamans like one with a "crooked finger" who famously left signature red handprints on one of the walls. As sacred sites, the caves may have been regarded with fear and avoided by the average paleolithic person. Likewise, in *Into Eternity*'s opening sequences director Michael Madsen strikes a match and speaks to us from the darkness inside Onkalo's human-created caves. Engineered by a new technocratic elite to house radioactive material with the capacity to mutate DNA and deform (or prevent) future life, Onkalo means "hiding-place." If it endures for the 100,000 years of its design specifications, this monument to what Madsen calls our "potent" society will be visited by no one.

Timefulness and Environmental Justice

So far, I have focused on the shortcomings of techno-managerial visions of the Anthropocene put forward primarily by scientists or science writers, especially a tendency of such thinkers to universalize the Anthropos and to avoid environmental justice issues. But I have also suggested there are merits to such accounts, particularly insofar as they can help us to re-think the temporal implications of "our" new geological-scale agency on the Earth. Is there a contradiction here? Certainly, but one that underscores deeper conceptual challenges at work in Anthropocene theory more generally. In their introduction to *Anthropocene Reading: Literary History in Geologic Times*, Tobias Menely and Jesse Oak Taylor usefully delineate the disciplinary roots of this apparent contradiction. After surveying humanist critiques of Anthropocene universalism, they observe that:

> [w]hat such criticism overlooks ... is that for scientists the designation of a single species as an agent is a *specifying* move rather than a universalizing one. The point is not that all humans are transforming the Earth system but that a single species in the biosphere is transforming the planet, a significant event in geologic time. The [Anthropocene Working Group] attempts to elide the problem of the social through its invocation of a stratigraphic signatory, while a critical humanities perspective insists on social variation and relations of power but is often inattentive to the broader biogeophysical systems in which humans intervene as a distinct agent. The perceived incompatibility between these positions on the Anthropos attests finally to the deep epistemological challenge of conceptualizing the double internality.
>
> (9)

There is much to unpack here, but the basic point is that scientists who address the Anthropocene are typically interested in what has become distinctive about humanity's species-level interventions on the planet as opposed to the agency of other species. This is why the Anthropocene Working Group (AWG), a panel of 33 scientists who advocated for the recognition of the Anthropocene as a new epoch, focused on identifying a geological marker (stratigraphic trace) recorded in the Earth as evidence of large-scale, novel changes our species has introduced, such as radionuclides from atomic weapons, evidence of higher CO_2 concentrations in the Earth's atmosphere, microplastics and novel anthropogenic chemicals. Of interest to these scientists were the geological traces of our recent (post-WWII), cumulative species-level activities, not questions of deliberate agency, historical origins or differentiated responsibility that typically concern those working in the humanities. As it turned out, the International Commission on Stratigraphy (ICS), the body that governs the geologic timescale, recently declined to recognize the Anthropocene as a new geological era, which I will address in the conclusion to this introduction. For now, suffice it to say that Menely and Oak Taylor's argument about scientists and humanists being at cross-purposes in how to understand our contemporary environmental crisis illuminates the epistemological difficulty of accounting for what they term "double internality," a concept developed by Jason Moore in *Capitalism in the Web of Life* to describe the complex, interpenetrating agencies of human social praxis (such as economics) and more-than-human geological and biological forces.

"Agency" in the Anthropocene turns out to be a wicked problem involving at least three major components: *Homo sapiens' species-level* geological-scale force on the planet (cumulative agency); the *differential agency* of distinct human groups and actors (subject-level agency); and more-than-human actants as defined by Jane Bennett, Bruno Latour, Donna Haraway and others (encompassing both technosphere systems like Bennett's electrical grid and the organic sympoiesis of Haraway's "critters"). In "The Human Condition in the Anthropocene," historian Dipesh Chakrabarty usefully employs the terms "homo" and "*anthropos*" to designate what I have called "subject-level" versus "cumulative" agency, observing that

> climate change as epochal consciousness … is constituted around a split between the homo, humanity as a divided political subject, and the anthropos, collective and unintended forms of existence … [our] geological force, as a species, as part of the history of life on this planet.
>
> (174)

As Chakrabarty suggests, it is an open question as to whether we should use the term "agency" to describe the anthropos, as most of its markers—like climate change—are currently unintentional, often being unwanted side-effects of our technologies. There is a thing-like quality to this cumulative agency that is deeply unsettling. Nonetheless, having attained geological scale such "agency" has now entered the arena of deep time. That is to say, we are currently producing changes in the Earth that stretch far into the future, from atmospheric greenhouse gases that can continue to change the climate for 100,000 years to the aforementioned high-level nuclear waste that can take between 10,000 to 1 million years to decay to safe levels. In *The Long Thaw*, climatologist David Archer predicts that "human climate forcing has the potential to overwhelm … orbital climate forcing, taking control of the ice ages. Mankind is becoming a force in climate comparable to the orbital variations that drive the glacial cycles" (6). Likewise, given that "anthropos" is currently driving mass extinctions of non-human species at a level that is estimated to be 100 times the background rate, this will impact the evolution of life on the planet for millennia to come. As Chakrabarty notes, we can no longer "background" (179) inhuman geological and evolutionary processes because we have

> fallen into "deep" history, into deep geological time. This falling into "deep" history carries a certain shock of recognition—recognition of the otherness of the planet and its very large-scale spatial and temporal processes of which we have, unintentionally, become a part.
>
> (181)

It is as if our capitalist technosphere and Gaia have combined to create an uncanny Frankenstein's monster, a shadowy hyperobject that both *is* and *isn't* "us," something recognizably anthropogenic but also ghastly and out of control. Little wonder that ecomodernists urge us to "love our monsters" (to coin Latour's memorable injunction) by mobilizing technology in order to create a more sustainable, deliberately chosen nature—and in doing so, to transform anthropos from something analogous to a meteorite hitting the planet to something resembling a rational agent.

Yet, as I have argued, without a "homo-level" environmental justice perspective to guide such development ecomodernists are potentially throwing more fuel on the dumpster fire we've already created. Their approach is particularly risky given incredibly powerful emergent technologies that are being developed from nano and quantum machines to synthetic biology and geoengineering. As Christopher Preston outlines in *The Synthetic Age*, many of these new technologies could have even more long-lasting (and potentially deleterious) impacts upon our planet than current tech. One of the most critical challenges we face today is how to bridge anthropos- and homo-level agency such that human praxis can interface in a more ecologically responsible way with the Earth's dynamic more-than-human systems. Fortunately, many passionately committed scientists and humanists have already begun the hard work of re-conceiving politics from a deep-time perspective.

For example, geologist Marcia Bjornerud's remarkable book *Timefulness: How Thinking Like a Geologist Can Help Save the World* makes a compelling case for "develop[ing] the habit of timefulness—a clear-eyed view of our place in Time, both the past that came long before us and the future that will elapse without us" (17). Bjornerud takes issue with the 24-hour model of life on Earth in which all of human history transpires within the last second before midnight as a "wrong-headed, and even irresponsible, way to understand our place in Time" for the following reasons:

> For one thing, it suggests a degree of insignificance and disempowerment that not only is psychologically alienating but also allows us to ignore the magnitude of our effects on the planet in that last quarter second. And it denies our deep roots and permanent entanglement with Earth's history; our specific clan may not have shown up until just before the clock struck 12:00, but our extended family of living organisms has been around since at least 6 a.m. Finally, the analogy implies, apocalyptically, that there is no future—what happens after midnight?
>
> (16–17)

Bjornerud analyzes ways that modernity conceived of as a radical break from the past leaves us "stranded on the island of Now" (166) and how capitalist economics exacerbates this presentism, to the deficit of future generations:

> From a purely economic standpoint, we should invest in preventing future environmental problems as long as the future benefits are greater than the present costs—and every economic study of the expected effects of climate change indicates that any investment now will repay itself many-fold. The real problem is shifting the timeframe for economic decision-making from fiscal quarters to decades or longer.
>
> (166)

In contrast to this temporal myopia, Bjornerud urges that we cultivate a sense of "timefulness" that is "polytemporal," celebrating different experiences of time in living things (a human lifespan versus a 5,000-year-old bristlecone pine or a 700,000-year-old soil bacteria) and acknowledging a deep continuity between our past ancestry, present challenges and different futures that are still possible. She sees a close relationship between timefulness and environmental justice, advocating for multi-generational thinking in political policies such as "tax incentives and subsidies [that] would be rebalanced to reward long-term stewardship over short-term exploitation" and recognizing that "poverty and class-disparity of opportunity [are] problems with deep historical roots that cannot be solved without sustained commitment over commensurate timescales into the future" (173–4).

Suggestively, literary scholar Jeremy Davies draws similar conclusions to Bjornerud regarding ways a geologically informed concept of time can influence political engagement in *The Birth of the Anthropocene*. Rejecting the Romantic notion that understanding the brevity of human generations within yawning expanses of deep time necessarily leads to political detachment—an impulse to "ris[e] above the present emergency in a spirit of enlightened impartiality" (8)—Davies draws the *opposite* conclusion, namely that a "geohistorical perspective … means accepting one's place in the actually existing moment of geologic time," which currently involves coming to terms with a "time of transition between the Holocene and the Anthropocene" (194). Davies suggests that instead of settling for Holocene notions of environmental sustainability (which often end up supporting status quo economic practices), environmentalists should re-conceive ecological activism within the context of Native American thought regarding the next "seven generations" and beyond:

> To recognize that is to see a new purpose for environmentalism. Not to escape from the crisis of the Holocene into a world made indefinitely "sustainable" and thereby liberated from geohistory, but something more demanding and more serious: to live within the crisis, and to struggle to influence its course by working for the survival of complex, pluralistic ecosystems.
>
> (194)

Like Bjornerud, Davies sees a strong affinity between understanding our place in deep time, environmental justice and emergent Anthropocene activism. By requiring an "avowedly global perspective" the new geohistoricism concerns itself with "the transnational effects of unequal relations of production and resource extraction," perceiving its "center of gravity as lying … with mass-based environmental justice movements of the majority part of the world" (202) instead of with a fairly narrow economic elite from the global north.

In this regard, Davies acknowledges the pioneering work of thinkers such as Rob Nixon who, in *Slow Violence and the Environmentalism of the Poor*, focuses on the disproportionate suffering experienced by citizens of the global south in the climate-change era, as well as the representational challenges associated with making forms of "slow" rather than spectacular violence visible. Nixon defines "slow violence" as forms environmental harm that "occur[s]" gradually and out of sight, a violence of delayed destruction that is dispersed across time and space, an attritional violence that is typically not viewed as violence at all" (2). Examples of slow violence include intergenerational economic colonialism and its environmental and social costs (Nixon examines how these are manifested in Nigeria's fossil fuel industry), deforestation and its link with poverty/loss of soil fertility over time (in areas like Kenya) and long-term toxicity from disasters like Bhopal in India, all of which are "slowly unfolding environmental catastrophes [that] present formidable representational obstacles that can hinder our efforts to mobilize and act decisively" (2). Davies acknowledges ways that Nixon's work still resonates within Anthropocene studies, but also expands upon it. He suggests that, rather than being a problem primarily happening "off shore" in developing nations, slow violence has become truly global in the climate-change era such that "this so-called environmentalism of the poor might appropriately become the type of environmentalism most readily associated with the idea of the Anthropocene," insofar as it disproportionately affects subaltern citizens in the global north as well as those in the global south (203). Moreover, unlike Nixon's book, *The Birth of the Anthropocene* includes non-human species in the roster of lives affected by climate dysfunction and calls for a politics that acknowledges the power of more-than-human actors, asserting that "the Anthropocene means widening the focus of sociopolitical critique and working toward *an analysis of power relations between geophysical actors, both human and nonhuman*" (62).

Indeed, a quest for *multispecies* justice is also a hallmark of Anthropocene theory that engages with geological time. In the dual introductions to *Arts of Living on a Damaged Planet*, Anna Tsing, Elaine Gan, Heather Swanson and Nils Budandt develop the interrelated tropes of "monsters" and "ghosts" in order suggest both the physical and temporal entanglement of living things, and ways that "living arrangements that took millions of years to put into place are [currently] being undone in the blink of an eye" such that "the hubris of conquerors and corporations makes it uncertain what we can bequeath to our next generations, human and not human" (G1). These authors explore "monstrosity" not as individualistic modernity defines it (as horrifying anathemas), but rather as life's basic ontological condition insofar as most organisms (including humans) are holobionts, communities of interlinked lifeforms that evolved sympoietically over time. Moreover, such "monstrocity" also applies to "symbiotic entanglement [that works] across bodies" (M2). These networks of interconnection and interdependency mean that monsters reflect "the wonders of symbiosis" but are also vulnerable to "the threats of ecological disruption." Likewise, in contrast to modern "progress" narratives that encourage forgetting of historical social and environmental trauma, Anthropocene "ghosts" are reminders of what has been lost as well as symbols for what can endure against the odds. In a time of escalating ecocide, the authors of *Arts of Living on a Damaged Planet* seek a viable long-term alternative:

> The deep time of geology, climate and natural science is collapsing into the historical time of human technology. *Anthropos* has become an overwhelming force that can build and destroy, birth and kill all others on the planet. In the new histories and politics that we must form, we must share space with the ghostly contours of a stone, the radioactivity of a fingerprint, the eggs of a horseshoe crab, a wild bat pollinator, an absent wildflower in a meadow, a lichen on a tombstone, a tomato growing in an abandoned tire. It is these shared spaces, or what we called haunted landscapes, that relentlessly trouble the narratives of Progress.
>
> (G12)

The issues explored by contributors to this collection amply demonstrate a tight coupling between environmental justice and multispecies kinship from a deep-time perspective. For example, in a condensed version of her book-length study, Deborah Bird Rose employs the Aboriginal aesthetic of "shimmer" to the illuminate the pulse of ecological patterns of death and life in Northern Australia, the terrible colonialist genocide committed against indigenous peoples and attempts to eradicate animals deemed to be "pests," but also the remarkable brilliance and resilience of Aboriginal culture and of native wildlife. What emerges from Bird's analysis is a sense of how all beings in this region are deeply entangled in relationships between flora and fauna that have co-evolved over eons, intimacies that underscore the critical importance of pollinators such as flying foxes.

Last but not least, Donna Haraway's extraordinary *Staying with the Trouble: Making Kin in the Chthulucene* undertakes nothing less than building an alternative vision of the Anthropocene, one eschews "the managerial, technocratic, market-and-profit besotted, modernizing, and human-exceptionalist business-as-usual commitments of so much Anthropocene discourse" (49–50) and instead weaves together different strands of "SF" loosely defined as "science fiction, speculative fabulation, string figures, speculative feminism [and] science fact, so far" (2). Rejecting both "comic faith in technofixes" and "game over" defeatism in the face of Anthropocene challenges, the book urges readers to "stay with the trouble" by getting their hands dirty and engaging in situated material praxis:

> Staying with the trouble requires making oddkin; that is, we require each other in unexpected collaborations and combinations, in hot compost piles. We become-with each other or not at all. That kind of material semiotics is always situated, someplace and not noplace, entangled and worldly … Neither despair nor hope is tuned to the senses, to mindful matter, to material semiotics, to mortal earthlings in copresence.
>
> (4)

In the service of modelling how we "become-with each other or not at all," Haraway develops the concept of sympoiesis (making-with) by drawing upon insights from evolutionary biology (such as Lynn Margulis's work on symbiogenesis), mathematical string theory, examples of symbiosis in ecosystems, collaborative artworks (such as the Crochet Coral Reef project), Indigenous Knowledge Systems, and the history of grassroots activism in different locales. She examines how sympoiesis can serve as a basis for multispecies justice by examining the deep historical practice and cosmology of Navajo rug weaving, interdependencies between this craft and Navajo-Churro sheep husbandry, historical cross-species unravelling via the US government's repeated genocidal campaigns and the remarkable restoration of Navajo-Churro sheep and the renaissance of Navajo activism in the Black Mesa Water Coalition and other groups. No less astonishing are the ways that *Staying with the Trouble*'s speculative fabulation not only addresses but *enacts* its sympoietic philosophy. The book's concluding "Camille Stories" expand upon a project begun in a writing workshop wherein Haraway, the filmmaker Fabrizio

Terranova and the philosopher Vinciane Despret were asked to "fabulate a baby" and to explore the life of that child "through five generations" (134). Working together with others, she develops a vision of the deep future wherein "Communities of Compost" arise from the "ruined lands and waters" of the twenty-first century to form new cross-species kinship bonds (every child is assigned an animal symbiont at birth), patterns which help to reduce human population pressures "while developing multispecies environmental justice of myriad kinds" (8).

Despite similarities between work on environmental justice within the context of deep time and ecomodernism's injunction that we exercise greater foresight and deliberation in how our technologies are re-shaping nature, it would be a great error to conclude that the thinkers I have cited believe the Anthropocene will be an era in which our species fully "controls" nature via technology and re-makes it in our own image. In different ways, all of these thinkers acknowledge nature's ongoing "wildness" even in a future when human social systems, including the technosphere, have adjusted to interface in more responsible and inclusive ways. Bjornerud celebrates the "wild" (178) geological revolutions in the Earth's deep past and thinks of the future as being something analogous to a parent's "aspirational visions" for their child—but such "aspirations" never amount to *control* over free agents. Davies, Rose, Haraway and Tsing all emphasize that we must acknowledge and respect the agency of more-than human entities with whom human lives are entangled. Indeed, such non-human actors (whether they be organisms, ecosystems, geological actants, climate-related feedback loops, etc.) must be taken into account as we develop future technologies and social policies lest we avoid making the same mistakes of our mechanistic past. As Steven Vogel's *Thinking Like a Mall* underscores, there is always an irreducible element of "wildness" in human artifacts and technology, an interaction with elements we did not create and do not control:

> No human transformation of the world ever fully produces the exact results those humans who undertake that transformation intended, in part that is because here is no such thing as a *merely* human transformation of the world. All transformations involve interaction of all sorts of forces, and only some of them are human forces: gravity plays a role, as do heat and cold and sunlight, as do—always—other living organisms as well.
>
> (166)

This is precisely why continuing to use technologies that are capable (either directly or indirectly) of destabilizing large-scale planetary forces like climate will only make our world *more wild* in the future—perhaps to the point of cataclysmic, out-of control feedback loops—that will exacerbate rising temperature and sea levels, loss of fresh water, increased release of methane from melted permafrost and warming oceans. It is also why re-thinking human history and our future aspirations within the longer context of deep, geological-scale time is so critical. Relative to our species' recent arrival on the Earth, geological time is like a profound *wilderness* of inhuman forces over which humans have (historically) had no control. But now that this is changing, we need to re-imagine not only our technological and social systems, but also what we mean when we say we would like to have better "control" over our planet. Great care must be taken with the language we use when speaking about our future aspirations and in the stories we tell about these horizons.

Literature in the Age of Geohistory

My survey of significant scholarship that addresses links between deep time and environmental justice in the previous section is, of necessity, truncated and incomplete. Nonetheless, it signposts some key studies that have influenced my own aims in *Wild Anthropocene.* In particular, this book will explore literature's potential for fostering an understanding of things such as "timefulness" and our ongoing sympoietic evolution with other life on the planet. As someone who regularly teaches courses on climate change fiction (cli-fi) and Anthropocene literature, it has been essential to address what literature and other artforms can bring to the table when discussing topics that are often dominated by scientific communication in public discourse. Initially, in the early years of the twenty-first century, many climate scientists and humanists alike were resistant to seeing value in new literary genres like "cli-fi," the former fearing that any association of climate discourse with "fiction" would contribute to denialism[6] and the latter segregating cli-fi into the category of "genre fiction" rather than "serious" literature, as Amitav Ghosh discusses in *The Great Derangement.* Yet, more recently, there has been an explosion not only in the publication of literature that engages with Anthropocene-era concerns such as climate change, globalism and extinction, but also in criticism that explores the conceptual and representational challenges negotiated in such texts. The volume and quality of such work underscores the extent to which Anthropocene literature has become mainstream rather than marginalized, its fiction encompassing the full gamut of genres (from realism and magical realism to speculative and weird fiction) and its poetry consistently experimenting with new and exciting formal innovations. I believe this renaissance is being driven not only by an increasing sense of urgency for effective international action on issues like climate change, but also by the fact that literature can mobilize emotional, phenomenological, philosophical and aesthetic engagement with such issues in ways that scientific data alone cannot.

Much critical engagement with Anthropocene literature has addressed representational difficulties associated with bringing together the individual, inward-looking subjectivity associated with Romantic and post-Romantic art with overwhelming material phenomena—such as climate change—which unfold at vastly larger spatial and temporal scales. Timothy Morton develops the concept of "hyperobjects" to describe such phenomena, which are "nonlocal" and "involve profoundly different temporalities than the human-scale ones we are used to" (*Hyperobjects* 1), yet nonetheless actively shape individual lives (both human and more-than-human) while resisting correlationist reduction to mere objects of thought. Given representational conventions involved with the "realist" prose tradition, including the need to set a story in a specific time and place, and to focus on the psychology and development of individual human characters, how can such fiction hope to capture the experiential reality of hyperobjects such as geological time? As Morton points out, specific weather conditions (like the rain falling on one's head—p. 11) do not adequately translate "nonlocal" hyperobjects such as climate. The same can be said for non-localizable deep time; those fossilized dinosaur footprints left in mud some 65 million years ago at what is now Dinosaur State Park in Connecticut are a *trace* of deep time [the "hyperobject of evolution" (88)], not deep time itself.

Why does any of this matter to literature? It matters because, as Ghosh puts it, with the merger of human and geological history in the climate-change era, we encounter a profound scalar disjunction:

> Here, then, is another form of resistance, a scalar one, that the Anthropocene presents to the techniques that are most closely identified with the novel: its essence consists of phenomena that were long ago expelled from the territory of the novel—forces of unthinkable magnitude that create unbearably intimate connections over vast gaps in time and space.
>
> (63)

Similarly, in *Ecocriticism on the Edge*, Timothy Clark explores the political and affective implications of such scalar incommensurability. He describes the Anthropocene as "an emergent 'scale effect'" wherein "at a certain threshold, numerous human actions, insignificant in themselves (heating a house, clearing trees, flying between continents, forest management) come together to form a new, imponderable physical event, altering the basic ecological cycles of the planet" (72). Activities that, on an individual level, may be life-enhancing—such as having children, improving one's standard of living, or driving a car—take on nightmarish proportions when magnified on the scale of billions of people engaging in such practices. As Clark puts it,

> what is self-evident or rational at one scale may well be destructive or unjust at another. Hence, progressive arguments designed to affirm individual rights and help disseminate Western levels of prosperity may even resemble, on another scale, an insane plan to destroy the biosphere. Yet, for, say, any individual household or motorist a scale effect in their actions is invisible. It is not present in any phenomenon itself, but only in the contingency of how many other such phenomena there are, have been and will be, at even larger distances in space or time.
>
> (73)

Clark's "Level III" scale human agency is equivalent to Chakrabarty's "anthropos" (i.e. our cumulative, species level impacts) and, like a "hyperobject," radically resists representation: "Can the Leviathan of humanity en masse, as a geological force, be represented? No, at least not in the realist mode still dominant in the novel. Its effects are global and non-localizable." According to Clark, two notable effects follow from the disjunction between what Charkabarty would term "anthropos" versus "homo"-level agency. On an emotional level, scalar disjunction can result in what Clark calls "Anthropocene disorder": feelings of rage, despair or impotence when contemplating the gap between individual actions/choices and global "scale effects" (like climate change) that unfold like "slow-motion catastrophe[s]" over long periods of time and appear to underscore "the almost complete irrelevance of my own 'decision[s]' at the present time" (140–1). On a political level, he asserts that "the scale of the Anthropocene entails a disconcerting *de-politicization*" or a separation of "new and imponderable ethical and cognitive questions … from the issues of cultural identity, politics, recognition, and so-forth" that have hitherto been central to a "progressive understanding of recent human history" (13).

Although Morton and Clark's accounts of the Anthropocene's representational challenges are thought-provoking and illuminate geohistory's novel scalar affects, they have a tendency to overstate the phenomenological, political and aesthetic aporias that follow from such complexity. For example, although climate-change is a "non-localizable" phenomenon that cannot be adequately grasped on the basis of isolated weather events (like Morton's raindrops), the increasing prevalence of superstorms, massive wildfires, flooding events and long-term drought are all devastating weather *patterns*—symptoms of larger climate destabilization. For those experiencing this destabilization's destruction first-hand, violent weather is all too in-your-face and localizable, rather than phenomenologically "ungraspable" in an abstract, philosophical sense. Likewise, it is questionable whether "de-politicization" necessarily follows from the Anthropocene's scalar dilemmas as Clark describes them. In his reading of literary texts such as Raymond Carver's short story "Elephant," conflicting accounts of agency considered at different scales does not result in a "de-politicized" stalemate in which political choice becomes meaningless, but rather in a sense of conflicting possible political paths moving forward: a first scale perspective that would preserve a sense of human rights and values individual identity and a "third scale" perspective that would tend to "register a person primarily as a physical thing" and "could soon drift towards the eco-fascistic" (108). Clark admits that "the challenge must be to continue and deepen the critique of individualistic conceptions of identity, right, etc. enmeshed as they are in the slow-motion catastrophes of international capitalism, without letting such a critique become an implicit endorsement of alternative, latently eco-fascistic forms of social control" (109), but this is hardly a "de-politicized" response to scalar disjunction. Likewise, it is suggestive that most of the examples Morton and Clark give of avant-garde art capable of reflecting the Anthropocene's new scalar dilemmas are not literary, but instead drawn from fields such as visual and sound art, as if to imply that there may be something "inexorably" (182) limiting about literature versus other expressive artforms.

Contrary to Clark's conclusion that considering the Anthropocene's "scale effects" creates an unbridgeable rift between anthropos and homo-level agency, I see contemporary literature as opening up space to think about how these two domains interact, informing one another in ways that stand to have long-term impacts and, thus, heighten the importance of present-day choices, rather than suggesting "the almost complete irrelevance" of individual "'decision'[s] at the present time." Indeed, all of the literature I examine in this book negotiates the question of how to better integrate anthropos- versus homo-level politics, rendering the scalar conflicts Morton and Clark theorize into vivid, phenomenologically convincing and impactful imaginative forms that dramatize embodied agents' struggles to navigate a new, climate-altered world. As with other scholars who have explored literature's innovative responses to deep time such as Lynn Keller and David Farrier,[7] my answer to Menely and Oak's question regarding whether "literary reading [can] provide insight into the Anthropocene's paradoxical alignment of precarity and agency, political urgency and deep time" (*Anthropocene Reading* 12) is a resounding "yes!" insofar as poets and novelists have devised imaginative ways to address such challenges.

In order to make way for such new literary paradigms, it is necessary to evaluate aspects of Holocene-era thought that have become problematic in the age of geohistory. In Chapter 1 of this book, "Robinson Jeffers, Deep Time and the Geological Sublime," I address one such hurdle: namely, Romantic concepts of deep time that dualistically separate geological-scale change from human history in the name of minimizing the latter's significance. One of the Anthropocene's hallmarks is that "Nature" can no longer be seen as a domain that is separate or "safe" from the impacts of human agency and vice versa. My first chapter examines this important shift in geological deep time's philosophical and political implications as evidenced in Robinson Jeffers's oeuvre. I examine Jeffers as a transitional poet whose work, on the one hand, resonates with Anthropocene concerns (such as a "geophilic" understanding of stone's more-than-human agency and anxiety about climate change), but who is also deeply invested in a dualistic understanding of Nature as human civilization's "Other." Jeffers values deep time for the same reasons he celebrates wilderness: namely, for its sublime transcendence of human agency and as a means to psychically escape from the traumas of history. Thus, he fails to fully anticipate a future in which ongoing social injustice goes hand in hand with drastically destabilized ecosystems and massive reductions in biodiversity. I employ Kirk Glaser's concept of the "geological sublime" as a framework for critiquing Jeffers's "Inhumanist" poetics, arguing that his recourse to geological deep time as a means to escape from historical trauma compromises forward-looking elements in an ethos that aspires to de-center the human in politics.

Chapter 2, "Angular Unconformities: Environmental Justice in the Poetry of Don McKay and Dionne Brand," contrasts the dualism in Jeffers's poetics with deep time's non-dualistic meanings in the work of two contemporary Canadian poets. I explore how both McKay and Brand attempt to bring "anthropos"-level awareness regarding humanity's geological-scale force into conversation with "homo"-level concern for environmental and multispecies justice. Unlike Jeffers, the arc of McKay's ecological philosophy begins with a perspective that Sophia Forster characterizes as "comic anthropocentrism" and ends in one that celebrates our species' capacity to become "members of deep time" along with countless other life forms that have evolved on the planet. For McKay, geological deep time is a profoundly de-familiarizing force, one that challenges his early emphasis on place-based *oikeios* as a ground for ecological politics. As a queer woman of Afro-Caribbean descent, Dionne Brand's poetic engagements with deep time in turn challenge the universalism at work in both Jeffers's and McKay's geopoetry. Brand's verse reveals why it is critically important not to forget links between geological deep time and the history of racist colonialism—connections between extractive geological regimes and "homo"-level socio-ecological injustice. As Kathryn Yusoff's *A Billion Black Anthropocenes or None* pointedly demonstrates, Anthropocene theory that universalizes geology as a politically neutral knowledge system often elides this science's historical role in mining, the development of fossil-fuel based technology and the enslavement of non-Europeans within colonialist regimes. Thus, both Brand's poetics and Yusoff's theory underscore our need to come to terms with the history of social injustice in order to avoid perpetuating neocolonialist capitalism that reduces both people of color and more-than-human entities to exploitable "resources."

In Chapter 3, "Imagining Extinction in Deep Time," my focus shifts from literature that engages with stratigraphic (geological) markers of deep time to texts that explore biological deep time. Specifically, I address Anthropocene literature that reflects anxiety regarding the future of biodiversity within the context of what many have described as a sixth great extinction event unfolding today. As Rob Dunn's *A Natural History of the Future* discusses, the discovery of microbes in the Earth's crust which exist for millennia and make up an estimated 20 percent of our planet's biomass, along with the extremophile adaptations of many bacteria and archaea, suggest that what is at risk from anthropogenic extinction is not "life on Earth" per se but rather those lifeforms that are our closest evolutionary kin, i.e. eukaryotic animals, plants and fungi. Likewise, the Earth's geological record reflects the fact that there have been several large-scale extinction events in which a majority of the planet's life was rendered extinct, but that these die-offs created the opportunity for new life to evolve. Chapter 3 argues that the risk of such deep-time perspectives on organic life is similar to that entailed with framing human history within the expanses of geological time: namely, an Olympian detachment from the ecocide we are currently perpetrating upon our closest living relatives and/or a disavowal of the importance of present-day political agency within the scope of evolutionary time. Theorists of the Anthropocene such as David Farrier have reached another conclusion: namely, that it is vital to recognize that anthropogenic mass extinction may be one of our longest-lasting legacies, one that will continue to reduce biodiversity millions of years into the future if we fail to alter current practices that are diminishing the evolutionary tree. As with the poets discussed in the first two chapters, contemporary fiction writers have created texts that encourage readers to realize the turning point we have arrived at in regards to extinction and to imagine different ways of negotiating this crisis. These responses range from techno-managerial proposals such as de-extinction to those that explore marginalized ecoterrorism as a desperate but unsuccessful praxis, to texts that suggest the need to mainstream multispecies justice within contemporary politics as a more sustainable, collective strategy to mitigating this problem. I explore how James Bradley's *Ghost Species* deconstructs real life de-extinction schemes to combat climate change by restoring megafauna and tundra ecosystems along the lines currently being pursued in Russia's "Pleistocene Park" and elsewhere. Bradley's novel critiques such techno-managerial policies as a symptom of capitalism's commodification of ecological mourning—a substitutive logic critiqued by Catriona Mortimer-Sandilands and others advocating for queer eco-melancholia as an alternative praxis grounded in multispecies justice. Jeff VanderMeer's *Hummingbird Salamander*, in turn, examines the ways that eco-melancholia can spur political activism but also the risks of such a politics, particularly when practiced by isolated and economically privileged individuals who resort to ecoterrorism when their utopian hopes for the future vanish. The chapter concludes by examining Karen Russell's short story "The Ghost Birds" as a call for greater multispecies justice in the face of unfolding climate catastrophe. This text depicts two forms of "adaptation" to this crisis in a near-future dystopian America: the first being neoliberal technological innovation that preserves life for some humans but also perpetuates widening social inequity and cascading extinction, and the second that recovers a sense of how our species has historically adapted its practices to accommodate marginalized groups in the name of achieving a more just society. Russell's story is haunted by the ghosts of New Deal era social policies and of past ecological adaptations that once allowed human and more-than-human agents to coexist, traces of historical utopian potential buried in the dystopian present but that, if recovered, could act as a road map for future progressive politics.

Chapter 4, "Reproductive Rights in Deep Time," addresses biodiversity and the related issue of human population growth in the Anthropocene era. Given concern regarding our species' "geological scale" agency, it is not surprising that geometric population growth in the twentieth century has been linked to environmental problems by theorists studying the Anthropocene, including those arising from industrial farming (the so-called "Green Revolution") needed to support such growth. Industrial agriculture is water intensive and heavily reliant on petrochemicals, produces significant CO_2 emissions and, due to a focus on cultivating monocultures, has contributed to loss of habitat and biodiversity for non-human species. And agricultural impacts are only one of many ecological challenges that can be associated with population pressures. Hence, a once taboo subject in ecocritical studies has re-surfaced in work by such prominent thinkers as Donna Haraway, Timothy Clark, Hannes Bergthaller and Eva Horn, along with calls to address population pressures when formulating a new politics that includes multispecies justice issues. While I acknowledge the legitimate concerns expressed by such theorists regarding overpopulation, this complex issue is (once again) best conceived of as a set of socio-environmental challenges that cannot be adequately addressed by technological "solutions" alone, such as improved agricultural techniques or better birth control methods. Indeed, given the dark history of calls for "population control" being linked with far-right reactionary politics (and with racism, xenophobia and totalitarianism) extreme caution must be exercised when addressing population pressures on a global level.

Nevertheless, considering the challenge of world population growth in conjunction with women's reproductive rights can be a very illuminating way to bring together anthropos- and homo-level politics. In Chapter 4, I argue that a commitment to global improvements in women's rights and racial equity (*homo*-level issues) must be part of any progressive intervention strategy for addressing population challenges. In order to counteract universalizing discourses within Anthropocene theory which falsely suggest equal responsibility for and uniform solutions to problems such as overpopulation, nuanced attention must be paid to the impact of patriarchal reproductive policies in both the global north and south. In this chapter, I examine literary texts that illuminate connections between women's rights, structural racism, environmental justice and population concerns. First, I examine Margaret Atwood's ecofeminist exposure of links between totalitarian patriarchy, denial of women's reproductive rights and ecological collapse in *The Handmaid's Tale* and its sequel *The Testaments.* The chapter then contrasts Atwood's politics in these novels with Louise Erdrich's in *Future Home of the Living God.* Erdrich expands upon and radicalizes Atwood's critique by focusing on ways that women of color are disproportionately affected by totalitarian reproductive regimes via ideologies such as social Darwinism and eugenics. Her critique adds a deep time dimension to reproductive politics by not only examining how a near-future authoritarian government distorts evolutionary theory to support eugenics, but also how similar patriarchal regimes authorized the historical genocide of native peoples. Moreover, Erdrich also links authoritarian policing of reproduction to speciesism, attempts to compensate for unease about our close evolutionary kinship to other animals by engaging in political practices that attempt to dominate, control and/or disavow our connection with other lifeforms.

It is our interspecies entanglement with other beings—or more specifically, the joys but also the vulnerabilities attending such enmeshment—that are addressed in the final chapter of my book, "Viral Capitalism, Tangled Trees and the Future of Biodiversity." My working thesis in Chapter 5 is that Anthropocene-era literary and theoretical texts expose neoliberal capitalism's viral temporality and material praxis, as well its destructive denial of both the benefits and risks arising from multi-species entanglement. Drawing upon David Quammen's study of horizontal gene transmission in *The Tangled Tree: A Radical New History of Life* and Andreas Malm's ideas regarding "parasitic capital" in *Corona, Climate, Chronic Emergency: War Communism in the Twenty-First Century*, I analyze the precarities attending entanglement from the perspective of disease transmission and how the neoliberal conquest of time and space has exacerbated these vulnerabilities, creating conditions for future global pandemics like the Covid-19 outbreak in 2019–2020. I examine how late capitalism's erasure of historicism and foreclosure of the future in a narrow presentism focused on short-term profit constitutes a major challenge to climate-change reform, as depicted in Margaret Atwood's *MaddAddam* trilogy and in Paolo Bacigalupi's *The Windup Girl.* Moreover, although this economic system creates class-based suffering and the material conditions for disease pandemics in both Atwood's and Bacigalupi's novels, its social inequities are masked by ideologies of biological essentialism and neocolonialism that work against political activism. I contrast these texts' pessimistic and apocalyptic assessments of ecological politics with the more hopeful reading of environmental advocacy in Richard Powers's *The Overstory*. Although tragedy is one prominent branch in Powers's monumental ode to our arboreal kin, comic survival is an equally striking one, the two perspectives ultimately woven together in a complex narrative mesh. My reading explores two key dimensions of the novel that can be read as progressive: first, its illumination of human-tree entanglement both in the lives of individual characters and over the course of several generations challenges readers to advocate for a politics that honors this condition and will insure it continues in the future; and second, its dramatic depiction of the deep evolutionary dimensions of this entanglement liberates time from capitalist foreclosure, creating a sense of temporal wildness and uncertainty in our future eco-social horizons. As with all of the other literature analyzed in my book, technology is imagined as part of the solution to Anthropocene challenges in *The Overstory*, but is not (in itself) sufficient to bring about necessary social change. Rather, the novel suggests that only by bringing a wide spectrum of human experience to the table (scientific, aesthetic and spiritual traditions, as well as diverse identity politics) might we develop a new politics premised upon multispecies justice.

Coda: The Anthropocene Epoch is Dead. Long Live the Wild Anthropocene!

On March 4, 2024 the Subcommission of Quaternary Stratigraphy (SQS), a subset of the International Commission on Stratigraphy (ICS), the official body that governs the geological timescale, made an announcement that sent shock waves around the world: namely, it voted *not* to recognize the Anthropocene as a new geological epoch (in a 12–4 split). Despite the fact that several AWG (Anthropocene Working Group) members contested the vote's legitimacy on procedural grounds, the ICS and the International Union of Geological Sciences (IUGS) upheld the vote, and issued a joint public statement about the decision on March 20. Oddly, this statement gives no specific reasons for the proposal's rejection, beyond a sense that

> it is clear from comments that were made during the course of the discussion period that many were unconvinced by the arguments in the AWG proposal, and their misgivings are clearly reflected in the decisive nature of the voting outcome.
>
> (IUGS)

The ICS/ IUGS statement cites *general* "cause[s] for concern" about recognizing the Anthropocene (such as the relatively recent starting date in the 1950s and the difficulty of reducing anthropogenic changes on the planet to any single starting point in time), but does not specify which (if any) of these was decisive in the SQS's decision, nor what was "unconvincing" in the large body of data that the AWG have compiled over the past 14 years.

Given that there has long been disagreement about what the Anthropocene signifies, it is not surprising that public opinion about the March decision has been divided. Scientists who were part of the AWG such as paleobiologist Jan Zalasiewicz, Earth scientist Martin J. Head, geologists Colin Waters and Simon Turner, and integrative biologist Anthony Barnosky have been vocal critics. In an opinion piece titled "Despite the Official Vote, the Evidence of the Anthropocene is Clear," Barnosky and journalist Mary Ellen Hannibal accuse "the gatekeepers of geological time [of] disregarding evidence of hundreds of peer-reviewed studies, claiming that the Anthropocene is a relative term that can't be pinned down in time and space, precisely identified, or measured" (*Yale Environment 360* online). Literary scholar Eva Horn and geologist Michael Wagreich, founders of the Vienna Anthropocene Network (VAN), excoriate the IUGS decision as reflecting "the extent to which the geological sciences remain, in part, disconnected from reality, and the inability of their official institutions to deal with the *present* state of the planet" ("VAN Statement on the IUGS Anthropocene Decision" online). Many critics of the decision also share Horn and Wagreich's concern that it "might … encourage the lobbies of science sceptics and climate change denialists that are still around and in power."

However, other commentators downplay the decision's significance and some even celebrate it as an opportunity to re-define the Anthropocene to make it progressive and/or convincing from the standpoint of geological science. For example, in an op-ed for the *New York Times* titled "Scientists Just Gave Humanity an Overdue Reality Check. The World Will Be Better for It," Stephen Lezak, an Oxford–Cambridge researcher who studies climate change politics, lauds the IUGS decision because "the politics behind the Anthropocene label were rotten to begin with," citing the Anthropocene's universalism, anthropocentrism and dualism as symptoms of this. Lezak even asserts that IUGS recognition of the Anthropocene as a new epoch would have been an "unacceptable act of defeatism" insofar as the long duration of geological epochs would suggest "we won't be getting out of this mess anytime soon … foreclose[ing] on the possibility that the geological future might be better than the present." Likewise, in "Not Yet Anthropocene: What the Official Rejection of Earth's New Epoch Means for the Climate Discourse," environmental author and activist Lou Sherry concludes that the IUGS rejection illustrates that "the Anthropocene is fundamentally political: [a] term that has transcended scientific semantics to become a political battleground" and claims that re-defining the Anthropocene as a "geohistorical event" would help us "understand how our present global reality cannot be decoupled from the material histories of Western capitalism and European colonialism, which lie at the origins of our destructive modern practices." In advocating for re-conceiving the Anthropocene as an "event" rather than an "epoch," Sherry modifies Erle Ellis's widely publicized response to the IUGS decision, summarized in an article he wrote for *The Conversation* titled "The Anthropocene is Not an Epoch—But the Age of Humans is Most Definitely Underway." Therein, Ellis discusses the fact that he resigned from his 14-year tenure on the AWG in 2023 when the group decided to designate 1952 as the beginning of the Anthropocene based on the global marker of radioactive fallout from hydrogen bombs tests. He claims he quit because "tying the start of the human age to such a recent and devastating event … risked sowing confusion about the deep history of how humans are transforming the earth." Indeed, Ellis also asserts that "if there is one main reason why geologists rejected [the AWG] proposal, it is because its recent date and shallow depth are too narrow to encompass the deeper evidence of human-caused planetary change." Here, Ellis reiterates a well-known critique of the Anthropocene: namely, that human beings have been modifying biodiversity and the Earth's climate for thousands of years (via prehistoric megafauna extinction, then through agriculture, colonialism and fossil-fuel powered industries and so forth), so why locate the Anthropocene's "origin" at a single point shortly after WWII? As an alternative to this quandary, Ellis proposes that the Anthropocene be defined as a geological "event" rather than an "epoch" because "events" are "complex, transformative and ongoing" rather than marked by a single globally synchronous impact detectable via a "golden spike" in the Earth's crust. The Great Oxidation Event is a prime example of a geological "event" insofar as it unfolded over a large swath of time approximately 2.5 to 2.3 billion years ago when anaerobic lifeforms went extinct due to oxygen emitted by cyanobacteria that evolved the ability to photosynthesize.

Although all of the responses cited here register valid perspectives, I tend to side with those that are critical of the IUGS decision as a missed opportunity to recognize what is *novel* (not merely contiguous) about our current geohistorical dilemmas. Zalasiewicz, Turner, Waters and Head summarize this argument in "What the Anthropocene's Critics Overlook—and Why it Really Should be a New Geological Epoch" (*The Conversation* online). They cite Ellis's criticism that the AWG's proposed post-WWII starting date is too "shallow" and "narrow" to "encompass the deeper evidence of human-caused planetary change" as an argument that "completely misses the point" underscored in Crutzen's original rationale for identifying the Anthropocene as a new geological epoch: namely that the *speed* and *scale* of human environmental impacts increased dramatically after the Second World War due to a variety of factors, including (but not limited to) exponential leaps in technological development (the "Great Acceleration"), geometric population growth, increased fossil fuel production and ubiquitous related products such as non-biodegradable plastics in packaging and other consumer goods, neoliberal capitalism's global hegemony after the fall of the Soviet Union, etc. Zalasiewicz et al. show what is missing in Ellis's proposal that the Anthropocene be re-defined as an "event" that has unfolded over millennia, namely, "any sense of the quantified rate and magnitude of [anthropogenic] change," or a "Y-axis" to supplement the "X-axis" of time: "The Y-axis is what scientists use to show the magnitude of measurements such as temperature and mass. It's absolutely critical to get an objective, number-based understanding of what is really happening." It's not that past evidence of anthropogenic change on the planet is negligible, as it is important to understand the historical origins for phenomena that continue to impact our current crisis such as the agricultural and industrial revolutions, extractive colonialism and capitalism. It's that these things, plus exponential growth in related areas have recently resulted in changes that are of a different order of magnitude than anything in the past. Or, as Zalasiewicz et al. put it, when the "Y-axis" is added to graphs mapping the history of human intervention on the planet, "the speed and magnitude of recent change jumps out at you" and it is these "sharp upturns [that] are essentially Crutzen's Anthropocene." These authors point out that sharp increases are also reflected in "similar graphs [that] can be drawn for species extinction," and, I would add, in graphs mapping increases in carbon dioxide emissions, human population growth, fertilizer consumption and water use documented in Bonneuil and Fressoz's *The Shock of the Anthropocene* (10–11).

Indeed, Ellis's call to re-define the Anthropocene as an "event" that has been unfolding ever since *Homo sapiens* was first able to devise and employ tools as a survival strategy greatly diffuses the "shock" of the AWG's version of the Anthropocene. By framing the Anthropocene as a story of slowly developing technological progress (from the fashioning of stone tools to agriculture to the industrial and atomic eras), Ellis emphasizes that there is "nothing new about human alteration of Earth's ecology" ("Too Big for Nature" 25) or about the Anthropocene per se, and implies that, as in the past, we can engineer our way out of current environmental challenges. The main change he and other ecomodernists call for is that we become more self-conscious in our current and future alterations because "the fate of both humanity and nonhuman nature does not depend on sustaining natural ecosystems but on the most proactive human reshaping of nature ever in history" (26). Yet, beyond downplaying the risks of such "proactive" technologies to cause unanticipated harm, such a perspective also overlooks a more troubling phenomena suggested in the AWG's account of the Anthropocene: namely, that it is has been our rapid technological *successes*, combined with much slower economic, social and multispecies justice reforms that have (arguably) resulted in the cascading crises of our present moment. Over 200 years ago, the poet Percy Bysshe Shelley anticipates this quandary in his *Defence of Poetry* (1821) when he notes that in the wake of industrial and agricultural revolutions,

> the cultivation of those sciences which have enlarged the limits of the empire of man over the external world, has, for want of the poetical faculty, proportionately circumscribed those of the internal world; and man, having enslaved the elements, remains himself a slave.
>
> (502–3)

My book is an attempt to create a bridge between humanist critiques of the Anthropocene and scientific perspectives that preserve what is shocking, urgent and provocative about its discourse. Given that Ellis favors re-defining the Anthropocene as an "event" roughly contiguous with the long history of human technological intervention on the planet, his response to the ICS/IUGS's rejection of the new geological epoch was to rally around the slogan: "The Anthropocene Epoch is dead; long live the Anthropocene" (*Conversation* online). But in order to avoid repeating this history's mistakes and to make room for more agents (human and more-than-human) in what could be a truly revolutionary future, I would prefer a different one: The Anthropocene Epoch is Dead. Long Live the Wild Anthropocene!

Notes

1 For a discussion of debates regarding the cause of megafauna extinctions, see Elizabeth Kolbert's *The Sixth Extinction*, pp. 227–35.

2 The "Anthropocene" has been a controversial term since it was popularized by Paul Crutzen and Eugene Stoermer from the year 2000 onwards. Criticism has focused on the universalism of "Anthropos" as a driver of change, with many humanities scholars coining alternative labels for the new era which aim to be more specific in identifying key agents of change. For example, thinkers such as Jason W. Moore or Andreas Malm who identify capitalism as a major driver of our current environmental catastrophe have proposed "Capitalocene" as an alternative nomenclature; those who focus on colonialist capitalism's agency such as Anna Tsing have suggested "Plantationocene"; while still others like Donna Haraway have proposed "Chthulucene" as a way of underscoring multispecies, entangled forms of agency. In this book, I will use "Anthropocene" to denote the new, post-Holocene era we are entering not because I think it is the best of all possible terms, but because it is (currently) the most widely used by both scientists and humanists.

3 For an overview of ecomodernist philosophy and policy proposals, see "An Ecomodernist Manifesto" (online) by John Asafu-Adjaye et al. Therein, the authors lay out their arguments for "decoupling" human well-being from reliance on natural resources (p.17 and elsewhere) via technological innovation. They envision the creation of a "good Anthropocene" as an extension of modernization, and make broad (and debatable) claims based on technocratic assumptions. For example, the authors claim that "meaningful climate mitigation is fundamentally a technological [my emphasis] challenge [insofar as] dramatic limits to per capita global consumption would be insufficient to achieve significant climate mitigation" (21). Likewise, the manifesto advocates for a "decoupling [of] environmental impacts from economic outputs" which reads like a justification for excluding serious critique of economic drivers for problems such as climate change.

4 Isabella Kaminski summarizes such concerns in her article "Why Geoengineering is a Human Rights Nightmare" (*The Wave* online).

5 See Elizabeth Kolbert, *Under a White Sky: The Nature of the Future*.

6 For an opinion piece that reflects such concerns, see George Marshall's "Climate Fiction Will Reinforce Existing Views" in the *New York Times* (online).

7 In *Recomposing Ecopoetics*, Lynn Keller explores how contemporary poets such as Juliana Spahr, Forrest Gander and Ed Roberson develop experimental forms to address what she calls the "scalar dissonance" of deep time. Likewise, David Farrier asserts that modernist poets such as Seamus Heaney and Elizabeth Bishop achieve a "thickening" of deep time (multi-temporal representations) via their aesthetic innovations in his book *Anthropocene Poetics*.

References

Archer, David. *The Long Thaw: How Humans are Changing the Next 100,000 Years of Earth's Climate*, Princeton University Press, 2009.

Asafu-Adjaye, John et al. An Ecomodernist Manifesto, April 2015, www.ecomodernism.org.

Atwood, Margaret. *The Handmaid's Tale*. Anchor Books, 2017.

Atwood, Margaret. *The MaddAddam Trilogy.* Anchor Books, 2014.
Atwood, Margaret. *The Testaments.* Nan A. Talese, 2019.
Bacigalupi, Paolo. *The Windup Girl.* Night Shade Books, 2010.
Bacon, Francis. *New Atlantis and The Great Instauration*, edited by Jerry Weinberger. 2nd edn., Wiley-Blackwell, 2016.
Barnosky, Anthony and Mary Ellen Hannibal. "Despite the Official Vote, the Evidence of the Anthropocene is Clear." *Yale Environment 360, Yale University*, April 2 2024, https://e360.yale.edu/features/anthropocene-denied.
Bonneuil, Christophe and Jean-Baptiste Fressoz. *The Shock of the Anthropocene: The Earth, History and Us.* Translated by David Fernbach, Verso, 2017.
Bjornerud, Marcia. *Timefulness: How Thinking Like a Geologist Can Help Save the World*, Princeton University Press, 2018.
Bradley, James. *Ghost Species.* Penguin, 2020.
Cave of Forgotten Dreams. Directed by Werner Herzog, IFC Films, 2011.
Chakrabarty, Dipesh. The Human Condition in the Anthropocene. *The Tanner Lectures in Human Values, Yale University*, February 18–19, 2015, https://tannerlectures.org/lectures/the-human-condition-in-the-anthropocene. Accessed December 2, 2024.
Clark, Timothy. *Ecocriticism on the Edge: The Anthropocene as a Threshold Concept.* Bloomsbury, 2015.
Crutzen, Paul. Geology of Mankind. *Nature*, Vol. 415, 2002, 23.
Davies, Jeremy. *The Birth of the Anthropocene.* University of California Press, 2016.
Dunn, Rob. *A Natural History of the Future: What the Laws of Biology Tell Us about the Destiny of the Human Species.* Basic Books, 2021.
Ellis, Erle C. "The Anthropocene is Not an Epoch—But the Age of Humans is Most Definitely Underway." *The Conversation*, March 5 2024, https://theconversation.com/the-anthropocene-is-not-an-epoch-but-the-age-of-humans-is-most-definitely-underway-224495
Ellis, Erle C. Too Big for Nature. In *After Preservation*, edited by Ben A. Minteer and Stephen J. Pyne. University of Chicago Press, 2015, pp. 24–31.
Erdrich, Louise. *Future Home of the Living God.* HarperCollins, 2017.
Fantasia. Directed by Ben Sharpsteen et al., Disney, 1940.
Farrier, David. *Anthropocene Poetics: Deep Time, Sacrifice Zones and Extinction.* University of Minnesota Press, 2019.
Ghosh, Amitav. *The Great Derangement: Climate Change and the Unthinkable.* University of Chicago Press, 2016.
Glaser, Kirk. Death, Desire and Domesticity in Jeffers's Pastorals of Apocalypse. In *Robinson Jeffers: Dimensions of a Poet*, edited by Robert Brophy, Fordham University Press, 1995, pp. 137–76.
Haraway, Donna. *Staying with the Trouble: Making Kin in the Chthulucene.* Duke University Press, 2016.
Horn, Eva and Michael Wagreich. "VAN Statement on the IUGS Anthropocene Decision." *University of Vienna Press*, April 22 2024, https://anthropocene.univie.ac.at/news-events/detail/news/van-statement-on-the-iugs-anthropocene-decision.
Into Eternity. Directed by Michael Madsen, Films Transit International, 2010.
IUGS. The Anthropocene. *International Union of Geological Sciences*, March 20, 2024, www.iugs.org/ _files/ugd/f1fc07_40d1a7ed58de458c9f8f24de5e739663.pdf?index=true.
Kaminski, Isabella. Why Geoengineering is a Human Rights Nightmare. *The Wave*, January 3, 2024, www.the-wave.net/geoengineering-human-rights-nightmare.
Keller, Lynn. *Recomposing Ecopoetics: North American Poetry of the Self-Conscious Anthropocene.* University of Virginia Press, 2017.

Kolbert, Elizabeth. *The Sixth Extinction: An Unnatural History*. Picador, 2014.
Kolbert, Elizabeth. *Under a White Sky: The Nature of the Future*. Crown, 2021.
Latour, Bruno. Love Your Monsters: Why We Must Care for our Technologies As We Do Our Children. *The Breakthrough Institute*, February 14 2012, https://thebreakthrough.org/journal/issue-2/love-your-monsters.
Lezak, Stephen. Scientists Just Gave Humanity an Overdue Reality Check. The World Will Be Better for It. *New York Times*, March 22, 2024, www.nytimes.com/2024/03/22/opinion/anthropocene-planet-geology.html.
Malm, Andreas. *Corona, Climate, Chronic Emergency: War Communism in the Twenty-First Century*. Verso, 2020.
Marris, Emma. Humility in the Anthropocene. In *After Preservation*, edited by Ben A. Minteer and Stephen J. Pyne. University of Chicago Press, 2015, pp. 41–9.
Marris, Emma. *Rambunctious Garden: Saving Nature in a Post-Wild World*. Bloomsbury, 2011.
Marshall, George. Climate Fiction Will Reinforce Existing Views. *New York Times*, July 29, 2014, www.nytimes.com/roomfordebate/2014/07. Accessed November 17, 2014.
McKibben, Bill. *The End of Nature*. [1989]. Random House, 2006.
Menely, Tobias and Jesse Oak Taylor, editors. Introduction. In *Anthropocene Reading: Literary History in Geologic Times*, Penn State University Press, 2017, pp. 1–24.
Moore, Jason. *Capitalism in the Web of Life: Ecology and the Accumulation of Capital*, Verso, 2015.
Morton, Timothy. *Hyperobjects: Philosophy and Ecology after the End of the World*. University of Minnesota Press, 2013.
Nixon, Rob. *Slow Violence and the Environmentalism of the Poor*. Harvard University Press, 2011.
Powers, Richard. *The Overstory*. Penguin Random House, 2018.
Preston, Christopher. *The Synthetic Age: Outdesigning Evolution, Resurrecting Species, and Reengineering our World*. MIT Press, 2018.
Quammen, David. *The Tangled Tree: A Radical New History of Life*. Simon & Schuster, 2018.
Russell, Karen. The Ghost Birds. *The New Yorker*, October 11, 2021, pp. 56–64.
Sherry, Lou. Not Yet Anthropocene: What the Official Rejection of Earth's New Epoch Means For the Climate Discourse." *Earth.Org*, May 6, 2024, https://earth.org/not-yet-anthropocene-what-the-official-rejection-of-the-anthropocene-as-earths-new-epoch-means-for-the-climate-discourse.
Tsing, Anna et al. *Arts of Living on a Damaged Planet*, University of Minnesota Press, 2017.
VanderMeer, Jeff. *Hummingbird Salamander*. Farrar, Straus and Giroux, 2021.
Vogel, Steven. *Thinking Like a Mall: Environmental Philosophy After the End of Nature*. MIT Press, 2016.
Yusoff, Kathryn. *A Billion Black Anthropocenes or None*. University of Minnesota Press, 2018.
Zalasiewicz, Jan et al. What the Anthropocene's Critics Overlook—and Why it Really Should be a New Geological Epoch. *The Conversation*, March 12, 2024, https://theconversation.com/what-the-anthropocenes-critics-overlook-and-why-it-really-should-be-a-new-geological-epoch-225493.

1 Robinson Jeffers, Deep Time and the Geological Sublime

There is nothing particularly new about literary representations of deep time that seek to counteract human hubris. Percy Bysshe Shelley's "Ozymandias," published in 1818 not long after Scottish natural philosopher James Hutton discovered deep geological time in 1789, famously critiques political tyrants such as Pharaoh Ramses II, whose historical legacy will gradually be erased over yawning eons. As an unnamed "Traveller from an antique land" testifies, little remains of a ruined statue to this leader, save a toppled head whose "shattered visage … and sneer of cold command, / Tell that its sculptor well those passions read" and "[t]wo vast and trunkless legs of stone" atop a pedestal that reads:

> 'My name is OZYMANDIAS, King of Kings.
> Look on my works ye Mighty, and despair!'
> No thing beside remains. Round the decay
> Of that Colossal Wreck, boundless and bare,
> The lone and level sands stretch far away.

Ironically, Ozymandias's "works" do, in fact, inspire "despair," but not of the sort he anticipated; instead of memorializing this monarch's unrivalled achievement, his monument reflects not only the cost in human lives required for his self-aggrandizement but also the folly of this project given the ravages of desertification and deep time (captured in the metaphor of "sands" that "stretch far away" and efface his statue). For philosophers such as Robert Frodeman, "Ozymandias" reflects the problematic conclusions humanists often draw about deep time's meaning, namely that "geologic time opposes human time, rather than encloses it—mocking our efforts … rather than being seen as part of them and ennobling them" (80). Frodeman's analysis raises fundamental questions about how we should think of human history within the much longer context of Earth history: does the great disparity between human and geological time suggest that political action in the present moment is (ultimately) of little importance within oceanic temporality, or should we draw the opposite conclusion—namely, that because what we do over the span of one generation can reverberate long into the future, the collective choices we make during our lifetime are of paramount importance? Contrary to Frodeman's reading, I believe both possible political readings of deep

DOI: 10.4324/9781003243946-2

time are at play in Shelley's "Ozymandias." Moreover, the questions that Shelley's poem raises about deep time's political implications have become all the more acute at the dawn of the "Anthropocene" era, now that humanity's collective socio-ecological impacts upon planet Earth have the scale and power of a geological force. If a Romantic conception of deep time is one that derives comfort from the thought that Earth forces dwarf and render impermanent all human endeavors, what happens to the political meaning of deep time when the power dynamic is reversed? What happens when, on a species level, humanity becomes a catastrophic form of *global* agency, capable of destabilizing the Earth's climate, of drastically reducing biodiversity and of producing long-lasting effects on the biosphere? As a thought experiment, imagine a shift from nineteenth-century industrial humanity's Ozymandian aspirations to lord over the planet and the weird situation we (inheritors of that era's revolutions) face today—a state in which we have collectively become something like the annihilating force of desertification itself, those "lone and level sands" which "stretch far away" in Shelley's poem. As David Farrier discusses in a 2016 *Atlantic* article titled "How the Concept of Deep Time is Changing," the Anthropocene condition greatly alters geological time's earlier Romantic associations, now that "the idea that humans are ephemeral compared to the workings of nature isn't as persuasive as it once was." What are deep time's political implications within this drastically altered state of affairs?

What has been called the "geological" turn in stratigraphic accounts of the Anthropocene is not only a response to our material (species-level) impacts upon the planet. Conceptually, it is also an attempt to forge a politics that is better suited to understanding the Earth's agency and relevance to human history than post-Enlightenment humanist frameworks. If humans have now become what Dipesh Chakrabarty terms "geological agents" capable of disrupting the Holocene's relatively stable Earth systems, phenomena such as extreme weather (superstorms, droughts lasting for decades, massive wildfires) and other unnatural disasters amply demonstrate that the world's ecosystems are *reactive* and increasingly chaotic. Hence new materialist thinkers such as Bruno Latour have identified the need to think "the Terrestrial as a new political actor." In the climate change era, the terrestrial is

> no longer the milieu or the background of human action. People generally talk about geopolitics as if the prefix "geo" merely designated the framework in which political action occurs. Yet what is changing is that, henceforth, "geo" designates an agent that participates fully in public life … [h]ow are we to distinguish from now on between physical geography and human geography?
>
> (41)

As Latour's analysis suggests, one risk of defining the Anthropocene as the "new Human era" is that the Earth's agency can be forgotten in calls to develop powerful new technologies (such as geo- and bio-engineering) as "solutions" to problems associated with climate change. Stratigraphic definitions of the Anthropocene retain a powerful sense of this agency by contextualizing human history within the longer reach of geological time. Deep time allows us to contemplate not only the long duration of phenomena which are the unforeseen by-products of our technology (radioactive isotopes, atmospheric CO_2) but also to imagine a time when our species may no longer exist—asking us to contemplate what ecological legacy we are leaving behind for the near and distant future. In other words, deep time in the Anthropocene is still a useful check on techno-managerial capitalism's Promethean aspirations, even given the historically unprecedented impact we are now having on the planet. But it is a way of thinking about deep time that must break with models that flourished during the Holocene. As recent books such as David Farrier's *Anthropocene Poetics* illustrate, studying differences in how pre- and post-climate-change-era poets imagine deep time can be indispensable in helping us to negotiate the transition from an older Holocene reality to our new Anthropocene one. In Chapters 1 and 2, I will address the political implications of this transition as illustrated in the "geopoetry"[1] of Robinson Jeffers, Don McKay and Dionne Brand. If critics such as Roy Scranton are correct in arguing that one of the challenges of our present moment is that we learn how to negotiate the "death" of our current civilization to make way for the future,[2] then it is instructive to examine ways that Jeffers's geopoetry is at once prophetic in analyzing the flaws in Holocene humanism and limited in its ability to grapple with forces that have (in the years following his death) become known as the Anthropocene.

Robinson Jeffers's Lithic Lyricism

Perched on a gently sloping knoll overlooking the Pacific Ocean and a much-travelled coastal road, there is a granite boulder which serves as the cornerstone for Tor House. Visitors to this famous stone dwelling, whose tower and garage were constructed by the poet Robinson Jeffers, are told this particular rock is named "Thuban." As James Karman explains in *Stones of the Sur*, Thuban is a star in the Drago constellation that, with Polaris, alternates as one of the Earth's polestars. Thuban was the polestar during the ancient Egyptian, Greek and Roman empires, but

> at about the time of Moses, Thuban … lost his place as lord of the night sky … after about twenty-five thousand years … Thuban will return and rule again. When this happens, it will be nothing new to Thuban the cornerstone; he has seen the whole wheel turn countless times before.
>
> (13)

Robinson and Una's propensity to name rocks like Thuban around their dwelling place might be easily chalked up to the eccentricities of artistic people with a Romantic bent. Indeed, until recently Jeffers was largely dismissed by scholars as a relatively minor "regional" poet, a writer whose verse and home shared the same dubious quality of fortress-like escapism from traumatic twentieth-century history. Although Jeffers's *Tamar* and other early volumes of poetry were critically acclaimed and propelled him into the public spotlight via *Times Magazine* coverage in 1932, his criticism of America's involvement in WWII and his ostensibly misanthropic "inhumanist" philosophy in *The Double Axe* (and other late poems) led to a sharp decline in his reputation within academia.

However, in the twenty-first century, something of a sea-change has occurred in Jeffers's critical reception, thanks in part to the publication of the five-volume Stanford University Press's *Collected Poetry of Robinson Jeffers* and three-volume *Collected Letters* of Robinson (and Una) Jeffers, as well as a growing appreciation of the poet's prophetic relevance to current environmental concerns. Ecocritical interpretations of Jeffers's verse, published in ShaunAnne Tangney's (2015) *The Wild That Attracts Us*, the journal *Jeffers Studies* and elsewhere, have addressed the poet's importance to place studies, ecofeminism and deep ecology, as well as his notable engagement with Earth science and astronomy. Such scholarship reveals the extent to which Jeffers anticipated ecological problems stemming from capitalist over-consumption, industrial harvesting and mass technics, geometric population growth and ubiquitous pollution. This is not to say there isn't lively debate among ecocritics regarding the political import of Jeffers's work. Some scholars, such as Bryon Williams, celebrate the place-based praxis of Jeffers's poiesis, while others like Temple Cone highlight what they see as a troubling indifference to human suffering in Jeffers's "inhumanist" philosophy that borders on ecofascism. While acknowledging the validity of these debates, I will argue that it is important to consider the issue of Jeffers's politics within an Anthropocene framework, particularly within the context of geological deep time. In order to better understand Jeffers's quite astonishing achievement as well as the political risks entailed in his "inhumanism," we must come to see ways his verse anticipates (and diverges from) what has come to be termed the "geological turn" in Anthropocene thought.

Within such a context, Jeffers's naming of "Thuban" is as much a reflection of his scientifically informed understanding of deep time as it is a mode of mysticism in the spirit of Thomas Aquinas's observation that stones "have something in them of the nature of stars" (quoted in Cohen 64). In this chapter, I have limited my discussion to Jeffers's fascination with geological deep time, bracketing his strong interest in astronomical deep time.[3] Specifically, I will examine two key dimensions in Jeffers's poetics that anticipate the current "geological" turn in Anthropocene theory—namely, his revelation of stone's vibrant materiality and his exploration of stratigraphic deep time—in order to build a case for the relevance of Jeffers's thought to what Latour (and others) call "geopolitics." As Jane Bennett notes in her essay "Earthling, Now and Forever?" "the idea of the Anthropocene tries to capture … a certain *convergence* between two styles of temporality we had formerly thought were distinct [the "geologic" and the "human"] … in order to challenge the assumption that "a hard, ontological dichotomy [exists] between human and non-human forces, with regard to their temporalities or scope of effect" (244–5).

I will argue that studying geological deep time's ethico-political import in Jeffers's verse is instructive because he is a transitional figure, one who has one foot in a Holocene worldview that valorizes nature's enduring imperturbability to (and transcendence of) human-produced change, and the other in an Anthropocene perspective that sees human and Earth history as inextricably entangled, more-than-human phenomena possessing an agency that modern culture ignores at its peril, even as humanity's capacity to destroy ecosystems has become more wide-reaching with each generation. Jeffers's "inhumanist" ethos is ahead of its time in calling for a thorough deconstruction of unthinking anthropocentrism, but his project of "un-centering" the human via what Kirk Glaser terms the "geological sublime" at times devolves into a misanthropic fantasy of wiping the Earth clean of our troublesome species so that Nature can rebound. The latter ironically ends up reinforcing a nature-culture dualism that is productively unsettled by other aspects of Jeffers's explorations of deep time. As Jeremy Davies argues in *The Birth of the Anthropocene*, in re-thinking human history within the context of geological time we should resist adopting "a distanced, Olympian perspective on the human condition" whereby a contrast is drawn "between the mere fleeting turbulence of humankind's concerns and the eternal currents of the great stream of life" (7–8). Davies argues that an alternative understanding of deep time actually works against such ahistorical detachment, yielding "above all a sense of [our] locatedness in time, a sense of being caught in one particular historical moment" (8)—a moment in which our collective actions matter as they will have material consequences that reach far into the future. My purpose in critiquing Jeffers's sometimes Olympian detachment at the prospect of human extinction is not to claim that those of us living in the Anthropocene era have moved beyond the temptation to find comfort in the thought of an "earth after us" (to cite the title of Jan Zalasiewicz's popular book). Rather, it is to illuminate ways that Jeffers's struggles continue to be our struggles, including the struggle to resist channeling rage and despair at ongoing eco-social devastation into misanthropic fantasies of human apocalypse. At its best, Jeffers's geopoetry is a testament to why what we do during our brief sojourn on this Earth matters and how we might live in ways that minimize our capacity for violence and destruction.

Embracing stone: Jeffers's Vital Materialism

Jeffrey Jerome Cohen's remarkable book *Stone: An Ecology of the Inhuman* (2015) explores rock's agency, ways this ostensibly mundane fellow time-traveler quietly solicits human speech. Possessing a "density, extensiveness, tempo, and force" that is alien to the individual human being's ephemeral existence, stone at once "rebukes epistemology" and is a "trigger to human creativity" (8). In particular, coming to terms with stone's capacity to endure over unimaginable millennia produces a vertiginous sense of "temporal noncoincidence" that, according to Cohen, "is profoundly, productively disorienting" (16). Although coming to terms with the fact that "stone's time is not ours" may be profoundly alienating, it may also inspire what Cohen terms "geophilia"—"a pull, a movement, and a conjoint creativity that breaches ontological distance," (19) via a "queer" intimacy that inspires thoughts of "the lithic in the creaturely and the lively in the stone" (20).

Though "love of stone" is the most direct etymological meaning of "geophilia," Cohen tries to avoid suggesting that this concept entails something like unilateral anthropomorphic projection (of human qualities or emotions onto stone). Instead, "geophilia" gestures at long-standing interrelations and interpenetrations between rock and human actants. For Cohen, "geophilia" is a powerful means to challenge anthropocentric exploitation of stone as "inert" matter or its reduction to a "natural resource" in extractive industries. Instead, "geophilia" entails an openness to stone's materiality, dynamism and even spiritual presence:

> The indifference of the earth to its dwellers paradoxically reinforces human exceptionalism, so that the material world comes to exist for our instruction and use. Yet stone refuses to remain fully set apart, to respect taxonomic distinctiveness. Because of its habit of undermining human singularity, of revealing common materiality as well as recurring affinity, to convey within its materiality the thickness of time, stone triggers the vertigo of inhuman scale, the discomfort of unfamiliar intimacy, and the unnatural desires that keep intermixing the discreet.
>
> (23–4)

Cohen claims that rock is "queerly productive" because its materiality "challenges as much as it fosters" those who engage with it, resisting Gaia's gendered, heteronormative and exploitable "fecundity." Importantly, as the "[m]onstrous child of the meeting of incompatible [temporal] scales," "geophilia" also entails re-imagining "ecology" as "Long Ecology" such that "choices being made in the present must be mapped backward in time to recognize the long histories they contain, and forward over the centuries of their potential unfolding" (41). Cohen's exploration of "geophilia" in *Stone: An Ecology of the Inhuman* therefore contributes to both the "new materialist" project of re-thinking things as something other than Cartesian objects and to Anthropocene discourse that seeks to re-conceive human history within the longer spans of geological time.

Indeed, Cohen's "geophilia" has strong affinities with Jane Bennett's concept of "thing-power" in *Vibrant Matter*, particularly insofar as Bennett's book explores mineral vitality as something that animates both geological and biological processes. As an example of this, she discusses Manuel De Landa's account of how, around 500 million years ago, bone first appeared in the bodies of animals, representing a "sudden *mineralization* … almost as if the mineral world that had served as a substratum for the emergence of biological creatures was reasserting itself" (quoted in Bennett 11, emphasis in original). Both Cohen's and Bennett's work radically challenges the post-Enlightenment assumption that stone, of all materials, most fully embodies matter's inert passivity. Turning this assumption on its head, Cohen's study dramatically illustrates how lively lithic presences have always enthralled our species, while Bennett's book explores the idea that humans "*are walking, talking minerals*" (Margulis and Sagan quoted in Bennett 11, emphasis in original). Significantly, both of these thinkers suggest that understanding how rock reflects "a vitality intrinsic to matter itself becomes more plausible if one takes a long view of time" (Bennett 10)—i.e. if one engages with deep, geological temporality.

Such claims for stone's vibrant materiality would come as no surprise to Robinson Jeffers. His lithic encounters reflect a thoroughgoing "geophilia" wherein he acted upon stone to build his dwelling-place but stone also profoundly *acted upon him*, transforming his verse via what Cohen might term a "queerly productive" intimacy. Many scholars have addressed how and why the period of Jeffers's stonemasonry represented a turning point in his growth as a poet. For example, in *Stones of the Sur*, Karman discusses how, in building Tor House "the ancient granite boulders, which [Jeffers] hauled from the shore and helped cement in place, opened up for him, revealing secrets about themselves, his own personality, and the cosmos as a whole." Karman cites a 1934 letter wherein Una Jeffers speculated that as her husband "helped the masons shift and place the wind and wave-worn granite" to build their home "he realized some kinship with it and became aware of strengths in himself unknown before" (cited in Karman 10). More recently, Bryon Williams's "Praxis, Gnosis, Poiesis: Inhabitation as Performative Myth in Thoreau and Jeffers" reveals a complex interrelationship between the knowledge Jeffers gained through the praxis of his stone-work, his artistic development and his quest to "truly inhabit" rather than "simply occupy" (164) the place he called home. While scholarship attesting to stone's crucial influence upon Jeffers's verse and experience of poetic dwelling is very important, I would like to push this further in order to illustrate a "geophilic" fascination with stone *as matter* that's also evident in his work. My purpose in doing so is to illuminate what I believe is a more radical vein in Jeffers's stratigraphic poetics—namely, his attraction to forms of materiality that actively resist reduction to human-centered needs (such as artistic achievement or a sense of being at home in the world). In *Vibrant Matter*, Bennett discusses how what she terms "geo-affect or material vitality [is] a theory born of a methodological commitment to avoid anthropocentrism and biocentrism—or perhaps it is more accurate to say that it is born of an irrational love of matter" (61). While Jeffers may not have consciously experienced "geo-affect" as a kind of "sex appeal of the inorganic" (to cite Mario Perniola's memorable characterization), there's certainly evidence in his work of an *eros* for stone that goes well beyond its usefulness as a building material or its capacity to make him a more potent poet.

This geophilic excess is already apparent in early poems such as "To the Rock That Will Be a Cornerstone of the House."[4] This poem's opening line addresses the granite bolder Thuban, but describes it as an "old garden" of multi-colored lichen, immediately blurring the distinction between mineral and organic life. For the speaker, this rock is not only a vibrant, individual presence, but also has a discernable history. The text's stanzas imaginatively reconstruct two distinct (though interrelated) strata that comprise Thuban's past: first, traces of its relatively recent history (its encounters with Native Americans and hawks) and second, how it has been marked by its more ancient, geological origins. In recalling the ghosts of "brown people who have vanished" from the site where Thuban sits and who "[b]uilt fires beside [it] and nestled by [it] / Out of the ranging sea-wind" (*CP* 1: 11), the speaker both attests to the comfort standing stones like Thuban have provided to different generations of human beings seeking a dwelling place and recalls a troubling history of genocide that displaced indigenous peoples who left a midden on the grounds of what would become Tor House. Like the Esselen and Ohlone peoples who once lived in Carmel, the speaker reveres the rock that will become a cornerstone for his home and wishes to ritualistically honor it, both with a poem and with a libation of wine, milk and honey. With these offerings, he marks the resumption of an intimate relationship that has long been absent from the stone: "A hundred years, two hundred, / You have been dissevered from humanity … no one / Touched you with love, … Where now my hand lies." Indeed, there is an oddly erotic quality to the description of the stone "drinking" its libation that once again blurs the boundary between the inorganic and the organic, the physical and the spiritual:

> I did not dream the taste of wine could bind with granite,
> Nor honey and milk please you; but sweetly
> They mingle down the storm-worn cracks among the mosses,
> Interpenetrating the silent
> Wing-prints of ancient weathers long at peace, and the older
> Scars of primal fire …

There are echoes here of the *Song of Solomon* 4:11 wherein Solomon admires his beloved: "your lips drip nectar, my bride; honey and milk are under your tongue," as the stone and the speaker are united via a queer communion that combines the lithic with vital bodily fluids (wine as symbolic blood mixed with milk). Yet, simultaneously, Thuban's origin and age are quite alien to the speaker, who observes that it possesses an "[e]ndurance that [has been] waiting millions of years to carry / A corner of the house." Though he often resorts to thinking about Thuban in terms of its use-value to him as a foundation for his home or a "comrade" for his old age, there is something in the stone's vibrant matter that also exceeds such utility, producing in him "the discomfort of unfamiliar intimacy" with a substance that "challenges as much as [it] fosters" (Cohen 24).

As in "To the Rock That Will Be a Cornerstone of the House," many of Jeffers's geophilic poems dramatize his physical and intellectual attraction to stone, but also ways in which its "otherness" resists his quest for intimacy. For example, "Tor House" (*CP* 1: 408) celebrates the fact that Jeffers developed "the art / To make stone love stone" via his masonry, but also contemplates a time "ten thousand years" into future when no "remnant" of his home will remain, outlasted by "the granite knoll on the granite / And lava tongue in the midst of the bay." Likewise, though poems like "Sign-Post" (*CP* 2: 418) urge those alienated by modern civilization to "[l]ean on the silent rock until you feel its divinity" as a means "to become human" again, other texts such as "Wise Men in Their Bad Hours" (*CP* 1: 10) describe mountains as "dead stone" whose "insolent quietness'" is as likely to inspire "hate" as "admiration" in the general public. This alternation between identification with stone and respect for its radical difference is in keeping with Jeffers's broader rejection of idealist understandings of nature. In "Credo" (*CP* 1: 239), he contrasts the philosophy of a friend "from Asia" who believes that "nothing [in nature] is real except as we make it" with his own "humbler" yet "harder mysticism":

> Multitude stands in my mind but I think that the ocean in the bone vault is only
> The bone vault's ocean: out there is the ocean's;
> The water is the water, the cliff is the rock, come shocks and flashes of reality.

A strange mineral alchemy unites the embodied human mind, housed in its "bone vault," with elemental ocean rock, but the speaker is careful not to conflate the ultimate "reality" of phenomena such as stone and ocean with what can be contained within the limited sphere of human perception. Like a bolt from the blue, rock "shocks" us with "flashes" of inhuman "reality" that cannot be reduced to any correspondence with human thought. Unlike an idealist, who would recoil at this "appalling" otherness, Jeffers finds in this very resistance a basis for love, celebrating rock's "heart-breaking beauty" as something that "was born before eyes" (human or otherwise) ever existed and which "[w]ill remain when there is no heart to break for it."

It is also significant that the rock which inspires the poet's admiration here is part of a natural geological formation (a "cliff") rather than one that has been used as a building material for human purposes. Indeed, although Jeffers's geophilia is most obvious in poems that meditate on stone associated with Tor House, other texts depict extraordinary lithic encounters at a distance from his home. Collectively, these poems attest to the poet's fascination with stone as a material presence, a power that exceeds its use-value for human beings. "Oh Lovely Rock" (*CP* 2: 546–47) a poem written on the occasion of a 1936 camping trip a 49-year-old Jeffers took with his son Garth (aged 19) and Garth's friend, Lloyd Tevis, is a particularly striking example of this. In a 1928 letter,[5] Una Jeffers responded to a question about her husband's working habits with an observation that at least once every fortnight the couple would go "back in the hills and coast redwood cañons" where Robinson would "examine stones & geological formations closely" (*CL* 1: 772). Clearly, the Jefferses enjoyed exploring the inland wilderness of the Carmel/Big Sur region, and Robinson was interested in the area's geological history. In the 1930s, the US Civilian Conservation Corps constructed a network of access trails into what was then known as the Ventana Primitive Area (now part of the Ventana Wilderness). It was into this rugged terrain of steep cliffs covered with dense chaparral that Jeffers and his teenage companions ventured on their 1936 camping expedition, with the objective of "climb[ing] a difficult mt [mountain]" (*CL* 2: 599) as Una put it in a letter dated July 19. This "inaccessible" (*CL* 2: 602) mountain, also mentioned by Una in a letter from August 10, was Ventana Double Cone (el. 4853 ft), which the group approached by first following the Pine Ridge Trail along the Big Sur River (Jensen 175). The opening line of "Oh Lovely Rock" specifies that the camping party "stayed the night in the pathless gorge of

Ventana Creek," indicating that the group followed this tributary of the Big Sur "up the east fork," going off-trail to access remote country. The opening stanza's description of this "pathless gorge" echoes Wordsworth's famous "Ravine of Gondo" passage in Book Sixth of *The Prelude* (1805). The speaker in Jeffers's poem expresses the same dizzying sense of being surrounded by vertical "rock walls" crowned by "forest on forest above our heads" as the speaker in Wordsworth's text, who, from within the ravine's "narrow" and "gloomy" chasm perceives "[t]he immeasurable height / Of woods decaying, never to be decayed" (lines 556–7) above him. However, whereas Wordsworth ultimately sublimates the Gondo ravine's disorienting Alpine landscape into "Characters of the great apocalypse / The types and symbols of eternity" (570–1), Jeffers's gorge retains a naturalism and specificity in its description. This is already signaled in the first stanza's identification of native tree species in and around the gorge, such as madrones, redwoods and rare high-elevation "Santa Lucian firs that stare up the cataracts / Of slide-rock to the star-color precipices." Though Jeffers's detailed evocation of the Ventana Creek gorge could easily have become a flight into rarified spheres of Romantic imagination, the first line of stanza two situates readers firmly on the ground: "We lay on gravel and kept a little camp-fire for warmth." Here, as in other lithic poems, rock appears to anchor Jeffers in a terrestrial world that refuses to be sublimated into airy musings.

But the wondrous experience of stone that unfolds in the second stanza is no mere drama of matter resisting mind. A majority of scholars interpret this experience as mystical, reflecting how Jeffers "made geologic process … into one of the profoundest expressions of his pantheism" (17) to quote George Hart's essay "Seeing Rock for the First Time." But commentators such as Hart also acknowledge that the poem's "mysticism" is inspired by the call of a decidedly material presence: namely, the gorge's massive stone wall across from Jeffers's campsite. Illuminated by campfire light, this wall strikes the speaker as commonplace:

> it was the rock wall
> That fascinated my eyes and mind. Nothing strange: light-gray diorite with
>   two or three slanting seams in it,
> Smooth-polished by the endless attrition of slides and floods; no fern nor
>   lichen, pure naked rock …

Yet somehow this rock face is also newly astonishing, creating in the speaker a sense that it is "as if [he] were / [s]eeing rock for the first time":

> As if I were seeing through the flame lit
> surface into the real and bodily
> And living rock. Nothing strange … I cannot
> Tell you how strange: the silent passion, the deep nobility and childlike
> loveliness: this fate going on
> Outside our fates. It is here in the mountain like a grave smiling child …

Although the speaker's experience could be described as "mystical" or religious, I would like to suggest that it also reflects a textbook example of what I have elsewhere described as the phenomenology of *wonder*.[6] The speaker's halting diction in the lines cited above reflect his surprise at discovering a vibrant materiality in diorite, "the real and bodily / And living rock" he has hitherto not perceived in his everyday encounters with human-worked stone. Although superficially the granite wall may appear to be static and inorganic—stripped of life such as ferns and lichen "by the endless attrition of slides and floods"—as igneous rock it was born of fire millions of years ago in the Earth's molten crust, and continues to be shaped by natural forces such as plate tectonics and erosion. The rock's apparent stasis is (therefore) illusory, resulting from an enormous gulf between human and geological timescales which renders its changes invisible to our species in real time. Likewise, the speaker speculates that long after he and his son have died, the rock wall will remain, but it will be not be "passive"; rather, "the energies / That are its atoms will still be bearing the whole mountain above."

While there is an irreducible element of mystery in this lithic encounter, it is also one informed by Jeffers's understanding of science and a naturalistic pathos for the fragility of embodied beings. The speaker identifies the gorge rock as "diorite," using a precise geological term rather than the more generic word "granite" which is often used by lay people to identify igneous rock. Moreover, the forces of erosion that are said to have shaped the gorge also reflect Jeffers's knowledge of geology, and the poem as a whole could be understood as a meditation on the relative rates of entropy that characterize human lives, tree life (whose remains Jeffers uses as kindling) and the "life" of extremely hard rock. Emotionally, there's a tenderness that follows from the speaker's understanding of human life's brevity relative to the longer life of stone, analogous to the small campfire that creates of an island of light in a wider darkness. For a brief moment, the poet sees the illuminated "vertical face of the great gorge-wall" and the faces of the young men sleeping by the campfire as *one* face, marveling at how an inhuman force "is here in the mountain" but also has the "silent passion," "deep nobility" and "loveliness" of "a grave smiling child." As Hart observes, this identification is unusual in Jeffers's oeuvre, as the poet typically chastises human beings for focusing their affection exclusively on other human beings to the neglect of more-than-human entities. Yet here, "the speaker's love for his son is immediately, or simultaneously, transferred to the rock across the creek" (20). I would argue that such "transference" is possible in this poem due (in part) to the way the text brings geological time into a complex conversation with human time. On the one hand, these temporalities converge in the speaker's apprehension that both human beings and rocks are subject to dynamic change, neither enduring forever, which inspires pathos for the impermanence of all things. On the other, the timescale involved in the rock cycle of substances like diorite versus the average human lifespan (millions of years versus a century at best) is so vastly different that it highlights a particular urgency in human existence, given we have so little time to make what we do over the course of our lives matter. We tend to think that the choices we make today will matter only within the relatively brief context of human-scale history, which, if we are parents, only extends to the next generation of our children. While this perspective may be pragmatic in debates about political policy, the convergence of geological and human time in Anthropocene thought complicates this issue. Having established ways that "geophilic" elements in Jeffers's verse anticipate such a convergence, we must now turn to the political implications of deep time in his poetry.

A certain detachment: Jeffers's geopolitics in the context of deep time

Like many of his Romantic predecessors, Jeffers believed that although poetry may address immediate current events, its primary focus should be more durable things—the natural world and its endless cycles of destruction and renewal constituting the most compelling example of the latter. In prose statements such as "Poetry, Gongorism, and a Thousand Years" (*CP* 4: 422–7) Jeffers aspires to create verse modelled upon the endurance of stone. In contrast to transitory "Gongorism" shaped by the vicissitudes of momentary fashion, Jeffers claims that "great poetry is pointed at the future," and its authors should strive "to be understood a thousand years from now," focusing on "the more permanent aspects of things" (423) as their subject matter. Though this commitment to verse that evokes what Wordsworth characterized as "the beautiful and permanent forms of nature" (339) appears to be fully consistent with the Romantic project outlined in the "Preface" to *Lyrical Ballads*, Jeffers also diverges from this project insofar as he believes attention to nature's timeless presence should have the effect of challenging our anthropocentrism. In his original 1947 "Preface" to *The Double Axe and Other Poems* (*CP* 4: 418–21) Jeffers counsels his readers to "Turn outward from each other, so far as need and kindness permit, to the vast life and inexhaustible beauty beyond humanity" in order to restore "moral and vital sanity" (420) to a twentieth century ravaged by war, totalitarianism, genocide and increasingly dehumanizing mass technics. In this document, the poet specifies that near the end of WWI, he conceived of his "inhumanist" philosophy, which entails "a rational acceptance of the fact that mankind is neither central nor important in the universe … an attitude [that] is neither misanthropic nor pessimistic nor irreligious, though … it involves a certain detachment" (418). What, though, is the nature of this "detachment" that comes from contemplating nature's "inexhaustible" forms? How does such detachment avoid the pitfalls of misanthropy, pessimism about history, political quietism and a general lack of "kindness" towards one's fellow human beings for which Jeffers was (and continues to be) critiqued? It is tempting to reduce the poet's "inhumanist" credo to a form of "Romantic ideology" or a flight from historical trauma into nature's timeless rhythms as a means to escape from responsibility to engage with real world political struggles. After all, isn't Jeffers the intellectual forefather of a particular kind of "ecological nihilism" (212), as David Wallace-Wells recently put it in *The Uninhabitable Earth*, that has inspired Paul Kingsnorth and others in the Dark Mountain Project to advocate "withdrawal" as the best political response to challenges associated with climate change today? Ecocritics seeking to recover a sense of what might be valuable or prescient in Jeffers's "inhumanist" philosophy must grapple with the issue of the poet's troubling tendency to valorize not only the endurance of rock but also its disinterest towards our species' historical struggles.

Indeed, one of the most consistent features of Jeffers's work is the comfort it expresses at the thought that no matter how destructive humanity becomes, nature will always transcend our species, retaining an ontological distance from our actions that will endure unfazed long after we are gone. As I will argue, it is this very distance that is collapsing in the Anthropocene, suggesting how problematic aspects of Jeffers's "detachment" have become as a way of responding to the convergence of geological and human history. This is why many contemporary theorists such as Davies derive exactly the opposite political conclusion from reconceiving human events within the context of deep geological time. Rather than detachment, most express a greater sense of urgency (and long-term consequence) in how we respond to present day socio-ecological problems such as climate change. As Davies puts it,

> The way to achieve [a] concrete and specific sense of the present, and to escape the temptation to embrace airy Olympian futurology, is to pay as much attention to what is ending as to what is beginning ... what matters about the Anthropocene, ethically and politically, is the way in which it is diverging from the Holocene ... the problem of the twenty-first century is how to negotiate a transition between these epochs. Which of the emerging kinds of divergence from the patterns established in the Holocene is it most important to avoid? Which can most bearably be tolerated? Which should we desire to bring about? (148)

Studying deep time in Jeffers's poetics reveals the extent to which he is a transitional thinker, at once invested in a Holocene understanding of Nature as a transcendent domain that will always remain separate from humanity and (simultaneously) in an ethos that strives to deconstruct ("decenter") anthropocentric hubris in the name of a more holistic worldview that anticipates Anthropocene-era perspectives.

To date, scholars who have addressed geological deep time in Jeffers's poetics have diverged in their assessments of this work's political implications. For example, ShaunAnne Tangney lauds Jeffers's frequent geological references as something that reflects his commitment to an idea of "organic wholeness" that encompasses both nature and humanity. For her, the poet's passion for Earth science qualifies our ability to label him an "apocalyptic" thinker in a religious sense, as his references to geological catastrophism suggest not an end to all earthly things but rather cyclical patterns of destruction and renewal which de-center humanity and underscore the notion that poetry is "only part of [a] larger earth language" (9). Contrarily, Kirk Glaser asserts Jeffers's geological references are very much in keeping with a religious "apocalyptic" imagination whereby the poet "embraces and praises ... obliterating nature as the only ultimate truth he can attain" (147). According to Glaser, Jeffers styles himself as "a modern Jeremiah who prophesies upon what he sees as the ills of modern civilization" (141) and who welcomes the prospect an apocalyptic future in which the Earth will be wiped clean of humanity's deleterious influence. Thus, "death and destruction (of the human world) are reconfigured, by being aligned with natural geological processes … [as] forces that reduce human efforts to nothing in the flow of a cosmic time beyond historical time" (147). Glaser coins the term "geological sublime" to describe this facet of Jeffers's thought, a concept whose political implications Temple Cone explores in his article "One Temper with the Granite: The Troubling Achievement of Robinson Jeffers's Ecological Lyric." Cone argues that poems such as "Calm and Full the Ocean" (written during WWII) reflect Jeffers's embittered recourse to a "geological sublime [that] reduces the value of human *life*" (19) and a disturbing lack of compassion for human suffering. For him, Jeffers's "totalizing ecocentrism, viewed in light of world historical events, demonstrate[s] a misanthropic stance bordering on ecofascism" (21).

While all of this scholarship is useful for understanding political controversy surrounding Jeffers's "inhumanist" philosophy, none of it connects his writing with ideas of deep geological time that have become crucial to "stratigraphic" accounts of the Anthropocene. A broader understanding of Jeffers's engagements with geological deep time illustrates the extent to which his verse was ahead of its time in challenging a narrow understanding of human history which equates "progress" with the technological mastery of nature and/or with our ability to "control" complex ecosystems. Tangney is right to assert that Jeffers employed ideas associated with geological catastrophism as a means to re-think human history as part of a much longer *Earth history* that stretches millions of years into the past and eons into the future. Only by contextualizing human history within the scope of geological time do we come to realize the extent to which civilization has relied upon relatively predictable Earth systems. *All of human history* has unfolded during the Holocene's generally stable climate—the last 11,500 years contrasting with earlier geological periods sometimes marked by rapid, catastrophic change. Today, there is growing evidence that accelerating anthropogenic climate change is pushing the Earth in the direction of a new period of cataclysmic instability for human cultures and ecosystems alike. This is why Davies develops a "neocatastrophic" theory of the Anthropocene, applying this concept from the Earth sciences to our current crisis wherein "[h]uman societies [now] exert their influence on the planet and so provoke the latest twist in a chancy, surprise-filled geological history" (9). Even scholars like Cone acknowledge that Jeffers's verse often reflects a prescient "proto-Deep Ecological" (14) sense of the disastrous consequences of modern capitalism's assumption that cheap natural resources are (in principle) limitless. It is extraordinary that Jeffers identified a perfect storm of what he termed "infantile" anthropocentric hubris ("Preface" to *The Double Axe*), excessive consumerism ("The Broken Balance"), exploitative land development ("Carmel Point"), geometric population growth ("Post-Mortem") and ever-more ecologically destructive (if economically expedient) technology as factors which would combine to produce "inevitable mass-disasters" ("The Purse-Seine", *CP* 2: 517) affecting future generations.

On the other hand, scholars like Glaser and Cone are also right to critique Jeffers's often reactionary emotional responses to present and future anthropogenic disasters, especially the solace he derives from a conviction that Earth that will flourish once humanity is gone. Jeffers's engagements with geological deep time are again significant here, as the poet often identifies rock's temporal endurance as a paradigmatic example of the Earth's ability to transcend our species in the long run. As we shall see, however, the poet accomplishes such a "detached" perspective at a high political cost. Within an Anthropocene framework, such an outlook seems at once too pessimistic in underestimating (and possibly discouraging) political action in order to avoid future catastrophe, and too optimistic in its assessment that "Nature" would escape unscathed if our species became extinct. On the whole, what we see in Jeffers's poetic explorations of geological time is a conflict between a holistic perspective that perceives human and ecological catastrophe as inextricably linked versus a detached, dualistic outlook that portrays human destruction as ephemeral compared with the power and beauty of eternal Nature. Though he is never able to resolve this tension, the conflict itself has important implications for Anthropocene politics.

Surveying Jeffers's oeuvre, there are so many poems that evoke geological time as a succor to human destruction it is impossible to engage with them all here. However, the late poem "Carmel Point" (*CP* 3: 399) published in the aftermath of WWII, is a representative text. In this text, the speaker lauds "the extraordinary patience" of natural "things" in "the beautiful place" where he lives, a once pastoral ecosystem that has recently been "defaced with a crop of suburban houses." Instead of bemoaning the permanent loss of this beloved place, the speaker derives comfort from the thought that the developers' "intrusion" will, in the long run, prove ephemeral. Imagining the landscape's perspective towards its violation, he declares:

> Now the spoiler has come: does it care?
> Not faintly. It has all time. It knows the people are a tide
> That swells and in time will ebb, and all
> Their works dissolve. Meanwhile the image of the pristine beauty
> Lives in the very grain of the granite,
> Safe as the endless ocean that climbs our cliff.—As for us:
> We must uncenter our minds from ourselves;
> We must unhumanize our views a little, and become confident
> As the rock and ocean that we were made from.

The speaker evokes deep geological time that "lives in the very grain of the granite" in order to "uncenter" (challenge) the assumption that human developers have the inherent right to convert nature's "pristine beauty" into real estate. Within a deep-time framework embodied in the granite cliff's ancient rock, such intrusions appear at once hubristic and fleeting. While the speaker appears "confident" that human enterprises will ultimately "dissolve" leaving the planet to recover its "pristine beauty," there is, in fact, a subtle tension in the poem between a holistic understanding that humanity and nature are one interfaced system and a dualistic sense that our species is an anomaly, a destructive force that must be removed before nature can heal. On the one hand, humanity is naturalized as "tide" that ebbs and flows, resembling the cycles of the "endless ocean" surrounding Carmel, and even the suburban houses are likened to an agricultural "crop" rather than artificial structures built of concrete, asbestos and steel. Such similes, and the speaker's pastoral nostalgia for a time when earlier ranchers used the land only for pasturing horses or cows, suggests that humanity has the capacity to live in harmony with the natural world, and that it is only capitalist greed/ exploitation that has "spoiled" the area. On the other hand, the speaker looks forward to the prospect of distant future wherein *all* human influence disappears and Carmel will be restored to "pristine" wilderness, valorizing a dualistic (absolute) separation of nature from culture and deriving comfort from the idea that the granite cliff and the ocean will both be "safe" from our actions in the long run.

It is the speaker's confidence in the latter scenario that seems most dated from an Anthropocene perspective. Today, industrialized nations are dramatically altering the Earth's ecosystems, producing effects that will linger far into the future. For example, climate change is causing the world's oceans to become more acidic (which is killing coral reefs and making it difficult for sea life to excrete calcium carbonate to build shells), and these oceans are also polluted with enormous gyres of plastic waste that, when broken down into particulate form, has entered the food chain, affecting ocean life and human beings alike. The most recent 2023 IPCC (Intergovernmental Panel on Climate Change) report acknowledges that climate change is happening at a more rapid rate than previously expected and that

> already, with 1.1 degrees Celsius (2 degrees Fahrenheit) of global temperature rise, changes to the climate system that are unparalleled over centuries to millennia are now occurring in every region of the world, from rising sea levels to more extreme weather events to rapidly disappearing sea ice.
>
> (Boehm and Schumer, 2023)

Significantly, the IPCC gives deep-time-related statistics to support the claim that these changes are "unparalleled" changes at work, such as the fact that concentrations of carbon dioxide in the Earth's atmosphere are the highest they have been for 2 million years, ocean acidification is happening at the highest rate in 26,000 years and sea-level rise is occurring faster than it has in any century for the past 3,000 years. If nothing is done to slow or halt climate change, prognoses for our planet in the near future range from grim (from an anthropocentric perspective) to catastrophic (from a holistic perspective that encompasses all Earth systems). David Wallace-Wells's *The Uninhabitable Earth* paints a bracing portrait of the former, outlining in great detail a future of sea level rise, flooding, superstorms, drought, fresh-water shortages, massive wildfires, crop failure, air pollution, ocean acidification and collapsing sea-life populations—all of which will displace millions of human beings as climate refugees. Other authors, including many climate scientists, have explored more encompassing long-term environmental problems which anthropogenic climate change may be triggering. Some have registered concern that rising sea temperatures could eventually lead to "methane hydrate destabilization," which could trigger the release of vast methane reservoirs currently stored deep on the floors of the world's oceans. Given that methane is a greenhouse gas that is 30 times more potent than carbon dioxide, a massive introduction of this gas into the Earth's atmosphere (in addition to that already being released by melting permafrost) could trigger what climate scientists such as James Hansen have called the "Venus Syndrome," a runaway greenhouse effect that could "destroy all life on the planet, perhaps permanently" (236).

Although the current scientific consensus is that such an apocalyptic outcome is unlikely given the temperature rise required to trigger the aforementioned methane feedback loop, most climate scientists also acknowledge that such feedback systems are not currently well understood and are (therefore) difficult to predict. A recent group of 16 scientists who published a paper titled "Trajectories of the Earth System in the Anthropocene" in the *Proceedings of the National Academy of Sciences* outlined a growing consensus that if carbon emissions aren't cut back "self-reinforcing feedback systems" will produce a "hothouse" planet with a climate not seen on Earth in millions of years (Steffen et al., 2018). According to scientists such as David Archer, in such a climate the next ice age would (effectively) be cancelled, underscoring the fact that changes we are making to the Earth's atmosphere today will continue to have effects for hundreds of thousands of years into the future. Likewise, through a combination of environmental toxicity, climate change, habitat destruction and the spread of bioengineered monocultures, our species is greatly reducing biodiversity in what has been likened to a sixth major extinction event, the long-term effects of which will continue to reverberate in the years to come. Finally, artificially produced phenomena which humans have introduced into the environment—from plastics to radioactive isotopes—can take thousands of years to biodegrade or become inert. All of which is to say that the comfort pre-Anthropocenes might have derived at the thought that the Earth is simply too large, powerful and enduring to be greatly affected by humanity's collective impact now seems (at best) misplaced and at worst an ideological evasion of our need to take responsibility for the destabilizing forces that we have unleashed on the planet.

It would be unfair to fault authors such as Jeffers (who lived and wrote primarily before the 1950s) for not fully grasping the long-term consequences of anthropogenic ecological destruction, given that many scientists locate the "Great Acceleration" following WWII as the origin point for the phenomenon we now associate with the Anthropocene.[7] Nevertheless, in light of what has become our geological-scale ecological footprint, it is important to question the dualism that often enables Jeffers to separate human suffering from nature's suffering, the fate of our species from the fate of the planet that we and a myriad of other beings call home. Some of the poet's texts come close to acknowledging a deep connection between human social injustice and our destruction of the natural world, but hesitate to fully explore this link in favor of a defensive need to construct nature as a "protected" domain untainted by the stain of human violence. Despite Jeffers's claim that such separation results from a "certain detachment" towards our species' fate, one can often see his recourse to such dualism as a form of psychological distancing from potentially overwhelming and painful trauma. For example, the poem "Calm and Full the Ocean" (*CP* 3: 124–5), which was written during WWII, opens with a gut-wrenching account of the "unimaginable agonies" which "consume mankind" in war-torn Europe and Asia, "uncounted millions" being subjected to "years" of "pain, horror," "sick hatred," famine and violence. In stark contrast to this suffering, the speaker finds solace in the natural world's "sane and intact" seasons, which unfold as they always have "as if man's world were perfectly separate from nature's, private and mad." As in "Carmel Point," it is the poet's confidence that "nature" will ultimately be "safe" from the stain of human violence and injustice that seems dated in the climate-change era. It is precisely this kind of nature–culture dualism that no longer seems viable.

But in the next line of "Calm and Full the Ocean," the speaker admits that such a dualistic perspective may "not [be] true," that even "P-38s and the Flying Fortresses are as natural as horse-flies." One can see an opportunity here for the speaker to ponder a connection between patriarchal and/or totalitarian domination that resorts to violence as a means of social control (manifested in the creation of ever-more-destructive killing machines) and modern humanity's unthinking violence towards the natural world. Such a connection is implied, for example, in the "high explosives" that rip up the Earth to produce "fountains of dirt" even as they shred human bodies into "flesh and bone-splinters." But instead of pursuing such connections, the speaker turns away from the spectacle of human violence he conjures in the opening lines, retreating from despair at human history by regarding it from the comforting distance of geological time. Thus, he can conclude:

> It is only that man, his griefs and rages, are not what they seem to man, not great and shattering, but really
> Too small to produce any disturbance. This is good. This is the sanity, the mercy. It is true that the murdered
> Cities leave marks in the earth for a certain time, like fossil rain-prints in shale, equally beautiful.

Unlike Cone, I do not think that "geologic sublime" in this poem "reduces the value of human *life*" (19) in an ecofascistic manner, nor do I believe the text tries to "argue that humankind's separation from nature has resulted in the destruction of war" (20). Instead of a callous disregard for the value of human life, I believe the opening lines reflect the speaker's profound horror at the large-scale loss of life during the Second World War, a trauma he responds to ideologically—i.e. by attempting to minimize the long-term effects of this violence from the vantage point of geological deep time. This Olympian perspective provides the succor of a "certain distance" from historical trauma, such that the speaker can naturalize the trace of "murdered cities" into something analogous to "fossil rain-prints in shale" and even to aestheticize such marks as "beautiful." Such a perspective is analogous to one a geologist might have upon rock that bears the traces of a past cataclysm that has (over time) left pleasing patterns that are separable from the catastrophic events that originally produced them. While I share Cone's sense that the poem is politically problematic, I believe it is so not only because it aestheticizes human suffering from the vantage point of (privileged) distance, but also because it evokes geological time in order to suggest that our violence (towards other human beings and towards nature) is "too small to produce any disturbance" in the long run, the trace of our destruction leaving "marks in the earth for a certain time" but then passing, like a rainstorm. While it may be true, as Jan Zalasiewicz argues in *The Earth after Us*, that in 100 million years "the Earth's surface will have been wiped clean of human traces" (14), this does not mean that it is appropriate to adopt a "distanced" perspective on phenomena such as climate change. In fact, as Zalasiewicz notes at the end of this book, knowing the long-term consequences of our current actions lends a sense of urgency to the present historical moment:

> The earth … is by far the most intricate, the most subtle, the most complex and valuable object in space for many, many billions of miles in any direction. It would be not merely an Earthly disaster if its surface was converted to the kind of wasteland that appeared after the Permian–Triassic or Cretaceous–Tertiary boundary extinction events. It would be a cosmic tragedy, one in which the injuries sustained would not heal for millions of years. This is a denouement that we should strive to prevent, while we are still able to.
>
> (240)

From an Anthropocene vantage point, perhaps the ultimate irony of the comfort Jeffers derives from the thought of an Earth rid of our species is that the dualistic separation of nature and humanity underwriting this vision might impede our ability to imagine a more ecologically responsible way for our species to dwell on the planet, an outlook that has had deleterious effects historically. As William Cronon identified in "The Trouble with Wilderness," fetishizing "pristine" nature that is untouched by human beings doesn't give us a model for how we might cultivate a sense of greater responsibility for the urban places we (increasingly) call home in the modern era. The Anthropocene has made all too visible the fact that there is no "elsewhere" to which we can escape the socio-ecological consequences of late capitalist modernity.

As *The Uninhabitable Earth* vividly illustrates, the catastrophes already unfolding due to climate change should be understood as both ecological *and* social disasters. The 2023 IPCC report stresses that not taking effective action today to mitigate (and/or to better adapt) to climate change will "exacerbate inequity should we fail to ensure a just transition" (Boehm and Schumer, 2023). Stratigraphic understandings of the deep time give us the ability to foresee the long-term effects of this potential catastrophe, showing the extent to which anthropogenic traces of phenomena such as climate change would not disappear even if *Homo sapiens* became extinct. Perhaps from the vantage point of granite's geologically sublime timescales, Jeffers could be forgiven for reveling in the "dark glory" of a "final unridiculous peace" that may one day be restored to the planet "after the inquisitive [human] animal's / Amusements are quiet" ("Hooded Night," *CP* 2: 3). But in the more immediate temporal moment—which is the only one in which vulnerable creatures of the flesh can live—we need to negotiate a more lasting peace now, while we still can.

Notes

1 Like the poet Don McKay, I borrow this term originally coined by the geologist Harry Hess, who is cited in McKay's essay "Ediacaran and Anthropocene: Poetry as a Reader of Deep Time." Hess created the term "geopoetry" to inspire imaginative engagement with his ground-breaking work in plate tectonics during the 1960s, asking his audience to "suspend their disbelief … long enough" for his ideas "to catch on" (McKay 46).
2 See Roy Scranton's popular study *Learning to Die in the Anthropocene* (2015).
3 For insight regarding Jeffers's interest in cosmic (astronomical) time, see studies such as Anthony Lioi's "Knocking Our Heads to Pieces against the Night: Going Cosmic with Robinson Jeffers" in *The Wild That Attracts Us*, pp. 117–40 or Steven Chapman's "On the Question of Science in 'The Inhumanist'" in *Jeffers Studies*, Vol. 8, No..2, Fall 2004, pp. 31–60.
4 All poems cited appear in *The Collected Poetry of Robinson Jeffers,* edited by Tim Hunt, 5 vols. For in-text citations, I will use the following abbreviation method for where to find specific poems: *CP* (*Collected Poems*), volume and page number. Because poems in *The Collected Poetry* do not have line designations, I omit these from citations.
5 Quotations from Una Jeffers's letters are taken from *The Collected Letters of Robinson Jeffers, with Selected Letters of Una Jeffers*, edited by James Karman, 3 vols. For in-text citations, I will use *CL*, volume and page numbers.
6 In *The Ecology of Wonder in Romantic and Postmodern Literature*, I explore the eco-political and philosophical implications of wonder and contrast this aesthetic with the sublime.
7 In May 2019 scientists who were part of the Anthropocene Working Group voted to designate the mid-twentieth century as the beginning of the Anthropocene. See Meera Subramanian's article "Anthropocene Now" for more details.

References

Bennett, Jane. Earthling, Now and Forever? In *Making the Geologic Now: Responses to Material Conditions of Contemporary Life*, edited by Elizabeth Ellsworth and Jamie Kruse. Punctum Books, 2013. pp. 244–46.

Bennett, Jane. *Vibrant Matter: A Political Ecology of Things.* Duke University Press, 2010.

Boehm, Sophie and Clea Schumer. 10 Big Findings from the 2023 IPCC Report on Climate Change. *World Resources Institute*, March 20, 2023, www.wri.org/insights/2023-ipcc-ar6-synthesis-report-climate-change-findings.

Chakrabarty, Dipesh. The Climate of History: Four Theses. *Critical Inquiry*, Vol. 35, No. 2, 2009, pp. 197–222.

Chapman, Steven. On the Question of Science in "The Inhumanist." *Jeffers Studies*, Vol. 8, No. 2, 2004, pp. 31–60.

Cohen, Jeffrey Jerome. *Stone: An Ecology of the Inhuman.* University of Minnesota Press, 2015.

Cone, Temple. One Temper with the Granite: The Troubling Achievement of Robinson Jeffers's Ecological Lyric. *Jeffers Studies*, Vol. 10, No. 1, 2006, pp. 5–25.

Cronon, William. The Trouble with Wilderness. In *Uncommon Ground: Toward Reinventing Nature.* Norton, 1995, pp. 69–90.

Davies, Jeremy. *The Birth of the Anthropocene.* University of California Press, 2016.

Economides, Louise. *The Ecology of Wonder in Romantic and Postmodern Literature.* Palgrave Macmillan, 2016.

Farrier, David. *Anthropocene Poetics: Deep Time, Sacrifice Zones and Extinction.* University of Minnesota Press, 2019.

Farrier, David. How the Concept of Deep Time is Changing. *The Atlantic*, October 31, 2016, www.theatlantic.com/science/archive/2016/10/aeon-deep-time/505922.

Frodeman, Robert. *Geo-Logic: Breaking Ground between Philosophy and the Earth Sciences.* State University of New York Press, 2003.

Glaser, Kirk. Death, Desire and Domesticity in Jeffers's Pastorals of Apocalypse. In *Robinson Jeffers: Dimensions of a Poet*, edited by Robert Brophy. Fordham University Press, 1995, pp. 137–76.

Hart, George. Seeing Rock for the First Time: Varieties of Geological Experience in Jeffers, Rexroth and Snyder. *Jeffers Studies* Vol. 8, No. 1, 2004, pp. 17–29.

Hansen, James. *Storms of My Grandchildren: The Truth about the Coming Climate Catastrophe and Our Last Chance to Save Humanity.* Bloomsbury, 2009.

Jeffers, Robinson. *The Collected Poetry of Robinson Jeffers*, edited by Tim Hunt, Stanford University Press, 1988–2000. 5 vols.

Jeffers, Una [1936] *The Collected Letters of Robinson Jeffers with Selected Letters of Una Jeffers*, edited by James Karman, Stanford University Press, 2011. 3 vols.

Jensen, Dan. *Robinson Jeffers: Fire from Stone.* Kaweah Press, 2016.

Karman, James. *Stones of the Sur.* Stanford University Press, 2001.

Latour, Bruno. *Down to Earth: Politics and the New Climate Regime.* Polity, 2018.

Lioi, Anthony. Knocking Our Heads to Pieces against the Night: Going Cosmic with Robinson Jeffers. In *The Wild That Attracts Us: New Critical Essays on Robinson Jeffers*, edited by ShaunAnne Tangney, University of New Mexico Press, 2015, pp. 117–40.

Margulis, Lynn and Carl Sagan. *What is Life?* University of California Press, 2000.

McKay, Don. Ediacaran and Anthropocene: Poetry as a Reader of Deep Time. In *Making the Geologic Now: Responses to Material Conditions of Contemporary Life*, edited by Elizabeth Ellsworth and Jamie Kruse. Punctum Books, 2013, pp. 46–54.

Perniola, Mario. *The Sex Appeal of the Inorganic: Philosophies of Desire in the Modern World*, translated by Massimo Verdicchio. Continuum Press, 2004.

Scranton, Roy. *Learning to Die in the Anthropocene: Reflections on the End of a Civilization.* City Lights Books, 2015.

Shelley, Percy Bysshe. Ozymandias. In *The Complete Poetry of Percy Bysshe Shelley*, edited by Donald H. Reiman and Neil Fraistat, Vol. 3. Johns Hopkins University Press, 2012, p. 326.

Song of Solomon 4:11. *English Standard Bible Online.* www.biblestudytools.com/esv/song-of-Solomon/4.html.

Steffen, Will et al. Trajectories of the Earth System in the Anthropocene. *Proceedings of the National Academy of Sciences*, Vol. 115, No. 33, 2018, pp. 8252–9. www.pnas.org/content/115/33/8252.

Subramanian, Meera. Anthropocene Now: Influential Panel Votes to Recognize Earth's New Epoch. *Nature*, May 21, 2019, www.nature.com/articles/d41586-019-01641-5.

Tangney, ShaunAnne. A Divine Physical World: Catastrophic Geology and Jeffers Country. *Jeffers Studies*, Vol 8, No. 1, 2004, 5–15.

Tangney, ShaunAnne, editor. *The Wild That Attracts Us: New Critical Essays on Robinson Jeffers.* University of New Mexico Press, 2015.

Wallace-Wells, David. *The Uninhabitable Earth: Life after Warming*. Penguin Random House, 2019.

Williams, Bryon. Praxis, Gnosis, Poiesis: Inhabitation as Performative Myth in Thoreau and Jeffers. In *The Wild That Attracts Us: New Critical Essays on Robinson Jeffers*, edited by ShaunAnne Tangney. University of New Mexico Press, 2015, pp. 161–92.

Wordsworth, William. Preface to Lyrical Ballads. In *Norton Anthology of English Literature*, Vol. D, 11th edn., edited by Deirdre Lynch. Norton, 2018, pp. 337–49.

Wordsworth, William. *1805 Prelude Book Sixth. The Prelude 1799, 1805, 1850*, edited by Jonathan Wordsworth et al. Norton, 1979, pp. 186–224.

Zalasiewicz, Jan. *The Earth after Us: What Legacy Will Humans Leave in the Rocks?* Oxford University Press, 2008.

2 Angular Unconformities

Environmental Justice in the Poetry of Don McKay and Dionne Brand

Angular Unconformity I: Don McKay's *Oikeios* in the Context of Deep Time

Canadian poet Don McKay's geopoetic vision can be seen as a hopeful alternative to Jeffers's tragic sense that human extinction may be the best hope for our planet's future. Geological deep time "speaks" through McKay's verse in a manner that resonates with contemporary geopolitics, allowing him to strike a balance between what Sophia Forster terms the "comic anthropocentrism" of his early work and Jeffers's more radical Inhumanist reading of deep time as a force that renders human history insignificant ("Don McKay"). Indeed, a running concern throughout McKay's oeuvre is the question of how—given our species' history of violent linguistic and material appropriations of the more-than-human world—we might learn to dwell upon the Earth in ways that promote greater social and ecological justice. Although his early work focuses on the unavoidable human-centeredness entailed in our modes of dwelling, experiences of place and language usage, his more recent writing qualifies a complacent acceptance of anthropocentrism via deep time's defamiliarizing force. It is McKay's engagement with geological time *Strike/Slip* and *Paradoxides* that makes his poetry relevant to debates surrounding possible meanings of the Anthropocene, particularly his sustained demonstration of deep time's power to deconstruct human hubris in our modes of present (and future) dwelling. As he notes in his essay "Ediacaran and Anthropocene: Poetry as a Reader of Deep Time," re-conceiving human history within the wider context of geological time ironically "puts a crimp in [the] anthropocentrism" often evoked under the sign of the "Anthropocene" (53), particularly techno-managerial understandings of this term which characterize humanity as a "Master Species" poised to take control of (hitherto) natural processes such as climate and genetic evolution. For McKay, thoughtful engagement with deep time can inspire us to "realign our notion of temporal dwelling" such that we ideally strike a greater balance between human-centered needs and respect for the wider ecological community of which we are "members" rather than masters. While we may have arrived at a point wherein humanity's collective impact on the planet is now analogous to a geological force, deep time reminds us that our species will also eventually become part of the fossil record. This reality,

DOI: 10.4324/9781003243946-3

in turn, should guide how we perceive home-making in the present, given the long-lasting effects our mode of temporal dwelling will have on future generations of human and more-than-human beings. McKay's poetry illuminates how deep time's de-familiarizing power *heightens*, rather than lessens, the urgent need to re-conceive *oikeios* ("home-making") as a process wherein social and ecological justice are irreducibly entangled rather than separable.

The arc of McKay's thinking about "how to come to grips with the practice of nature poetry in a time of environmental crisis" (*Vis à Vis* 9) can be seen in the essay "Baler Twine." In this piece, the poet attempts to reconcile a tension in between humanity's need for home-making and what he terms "matérial," between the inescapable appropriation entailed in the former and "total appropriation" (19) involved in the latter. At the outset, McKay notes that the essay was prompted by his troubling encounter with a dead raven "hung up by the roadside at the entrance to a lane, a piece of baler twine around one leg, wings spread," a bird whose body had not only been shot but also put on display. There follows a meditation on the relationship between human dwelling (in both a phenomenological and poetic sense) and our drive to dominate/appropriate nature as "other." The poet speculates that two forms of appropriation were involved in the killing and exhibiting of the roadside corvid—forms of violence that function as analogues for "our relations to material existence" more generally. McKay associates the first appropriation with the familiar Heideggerian concept of "standing-reserve": the process of converting natural phenomena into tools that have use-value for human beings, but (in the process) are denied their autonomous existence. Although the poet does not speculate on the raven's use-value as "standing-reserve," it can be inferred that the bird has been reduced to a symbolic "tool," or a means towards different semiotic ends. Its utility functions on multiple levels: marking an individual's private property (capitalist signification), demarcating his or her superiority to the rest of the animal kingdom (speciesist ideology) and/or symbolizing nature's subjugation within modern agricultural praxis (baler twine binds the bird's corpse just as it can be used to bind hay or straw). Yet there is an *excess* in spectacle of the dead raven that seems to go beyond its reduction to utility, and it is this extra something that McKay calls "matérial," a second-order appropriation wherein human beings either *colonize* or *deny* (20) *death itself* via violence towards more-than-human others. As the poet surmises,

> Shooting the raven was one thing: we all know, each of us, that sinister delight in casual brutality and long-distance death. Displaying it was another—controlling its death, as well as taking its life. Displaying it declares that the appropriation is total. A dead body seeks to rejoin the elements; this one is required to function as a sign, a human category—a sign that simply says, "we can do this."
>
> (18–19)

The problem McKay struggles with in the essay is how to avoid matérial (and the militaristic/utilitarian violence this term implies) in our modes of human dwelling on the earth if some degree of "appropriation" is unavoidable whenever we establish living places and because we must perceive the natural world through human "linguistic and cultural categories … the mental equivalent of baler twine" (29). McKay sees home-making and poetic composition as parallel processes, each involving a drive to possess the other (the removal of things' autonomy via ownership and naming) *and* the potential for human subjects to open themselves to otherness (by paying attention to, appreciating and celebrating the material world).

Playfully anticipating a poststructuralist critique of his essay as the musings "Mr. Nature Poet, standing by the roadside, outfitted no doubt by L.L. Bean, happily twirling your dead raven [when] it's a fact that you're going to crash into language in about .05 seconds," McKay observes that

> [n]ature poetry's paradoxical situation is … analogous to home-making. Being language, it cannot avoid the primordial grasp, but this occurs simultaneously with the extended palm, the openness in knowing what I've been calling poetic attention … nature poetry should not be taken to be *avoiding* anthropocentrism, but to be enacting it, thoughtfully. It performs the translation which is at the heart of being human, the simultaneous grasp and gift of home-making.
>
> (29)

As an example of the kind of "thoughtful" anthropocentrism entailed in *oikeios* and poiesis as processes of "translation," McKay explores anthropomorphism in the penultimate section of the essay. In moments when "ownership is set aside" in our relations with the natural world, "appropriation can turn inside out" to create

> an opening, a way of going up to something with a gift from home. Growths on this stump remind you that the Japanese call certain fungi "tree ears" … you see ravens playing on the Athabaska River and think "boys on a raft." Anthropomorphic play, along this trail, is a gift to the other from the dwelling you will never build there.
>
> (31)

Although McKay claims that the reciprocal give-and-take involved in such anthropomorphic projection is "dangerous" because in it "language … compromises its nature, dismantling itself in a gesture towards wilderness" (31–2), the act of "translation" here is, in fact, quite safe insofar as it maintains a comfortable anthropocentric perspective. Thus, trees covered in fungi resemble human heads with cute "ears," rather than the far more unsettling prospect of the human body resembling tree–fungi systems insofar as our microbiome contains myriad symbiotic networks. Moreover, the "anthropomorphic play" or "erotic hinge of translation" McKay celebrates can unfold only in the absence of "ownership" and the kind of land development that's requisite for human dwelling in a *material* sense. He has already shown us ways such ownership and material use can degrade into "matérial" which makes a spectacle of murdering more-than-human others (such as the baler-twined corvid), a problem that isn't solved by the poetic "gift" of imaginative projection when we anthropomorphize such others (i.e. our ability to see crows as analogous to "boys" playing on a river). To McKay's credit, in the final section of "Baler Twine" he acknowledges the limits of poetic dwelling and mourning rituals as things which mitigate against matérial, stating

> I doubt if these work with matérial, where the death-and-reduction comes thoughtlessly, from long distance … [t]here is no ritual imaginable which would, right now, set in balance our relation to Pacific atolls blown up in hydrogen bomb tests, or to clear-cut forests, or to the ecosphere itself.
>
> (33)

Sophia Forster sees a deep affinity between Joseph Meeker's definition of "comic" environmentalism and McKay's humility regarding the scope of poetry's power, his verse's experimental linguistic play and openness to contingency. She employs Meeker's well-known distinction between "comic" and "tragic" environmental philosophy in *The Comedy of Survival* to build a case for reading McKay as a poet with a poststructuralist interest in language's mediating role when we interface with the more-than-human world. She contends that this places McKay outside of the dominant "mimetic" tradition of nature writing and ecocriticism, and lauds his ethos of "thoughtful" anthropocentrism as a non-tragic disavowal of linguistic/epistemological mastery. Yet, despite highlighting important dimensions in McKay's philosophy, Forster's conclusion that his verse is "useless … in the most necessary ways" in its "refusal to submit to an instrumentalist view of language or the world" doesn't adequately address the poet's oft-repeated concern about how to write about "nature" in a time of environmental crisis. I believe McKay's more recent work deconstructs the limits of "thoughtful" anthropocentrism as a basis for benign home-making. As I will illustrate, the geological turn in McKay's poetry has enabled him to engage more effectively with the problem of matérial and human *oikeios* within the context of deep time.

Suggestively, it was McKay's immersion in a landscape not unlike the Ventana Creek wilderness where Jeffers camped in the 1930s that contributed to this upheaval in his thinking. In 1996, he moved to southern Vancouver Island in order to explore the Loss Creek–Leech River Fault, a dramatic, steep-sided canyon which (like the Vantana Creek gorge) was created by strike–slip plate tectonics. In "From Here to Infinity (or So)," McKay reminisces about his reasons for exploring this landform, such as wanting to get to know this place from the "bottom-up, rather than the top-down" (120), and, consequently, not only walking the entire length of the fault, but also educating himself about its geology and its history of human inhabitation, from indigenous cultures to more recent European exploitation during a gold rush and (later) as a site of "severe" clear-cutting (119). Rather than simply providing a nuanced experience of place, however, McKay discovered that his Vancouver immersion brings into focus "the relation of hereness to deep time and the infinite" (120), an experience that is at once exhilarating and disorienting:

> When I got hooked on the Loss Creek project, I thought I would be improving my sense of place. But I found that my sense of place was itself destabilized by the addition of deep time and plate tectonics. It was enriched, certainly, but enriched in such a way that threatened the idea of place itself. Such is the effect of infinitosis, that perilous and necessary disorder and, of course, I could have encountered it anywhere on the planet by picking up an "ordinary" pebble and enquiring of its origins … But I'm not alone, I'm guessing, in requiring such gross stimuli as a beautiful, forlorn fault to dislodge me from my comfortable assumptions about place and time.
>
> (122–3)

Extending Jorge Luis Borges's idea that the infinite is a "concept which corrupts and upsets all others" (Borges 202), McKay coins the humorous term "infinitosis" to denote the perilous (yet also necessary) effect of our encounters with phenomena of such immense size and scope that they dwarf human enterprises and threaten our ability to imagine them adequately. In a cheeky nod to this experience's Romantic antecedents, the poet notes that "a mild, or homeopathic, dose of the infinite is the crucial element in the aesthetic experience known as the sublime" (116); as we shall see, however, the effects of McKay's poetic encounters with deep time also diverge significantly from the Romantic/Kantian sublime insofar as they focus (rather than transcend) attention to problems stemming from eco-social injustice.

In "From Here to Infinity (or So)," McKay discusses how the rock in Loss Creek canyon made the relation "of hereness to deep time and the infinite" quite vivid because the fault lies on the surface rather than under yards of overburden, such that

> For most of its length, you have the layered sedimentary beds of the Pacific Rim Terrace on the one side and, on the other, a fist of igneous basalt … What happened forty-five million years ago is that this igneous fist, having been carried northeastward on the Pacific Plate, slowly crashed into its predecessor.
>
> (120)

To put a number like "forty-five million years" in perspective, this is over seven times the amount of time since our ape-like ancestors first evolved (about 6 million years ago); massively longer than the time when *Homo sapiens* first appeared on the earth (around 200,000 years ago) or the beginning of civilization as we know it (about 6,000 years ago), and mind-bogglingly longer than the average global life expectancy (approx. 73 years in 2024).[1] Attempting to come to terms with such enormous duration from the vantage point of human time is, therefore, risky. As McKay notes, this can have the effect of inspiring one to "give up and collapse into a reverie in which your life grows tinier and tinier, and its details … grow less and less consequential" (117). Except that today—as the poet is well aware—we have arrived at a point when our *collective* effects on the planet are no longer "tiny" either in scope or duration.

Although McKay doesn't explicitly connect the need to re-think "hereness" with our Anthropocene condition in this essay, his insistence that "the idea of infinity (or its slightly less perilous cousin deep time)" is a *necessary* component in how we re-think the issue of place strongly suggests that this is because such thought is not only "an antidote to human hubris" which demonstrates that "man is not the measure of all things," but also because such thought can inspire us to think more rigorously about the long-term effects of our current actions. According to McKay, poetic metaphor (rather than abstract number) enables us to *feel* deep time intensively and can significantly impact how we choose to dwell upon the earth:

> One metaphor often used for the effect of metaphor in relation to deep time or infinity is to say it "brings it home." That is ... metaphor brings infinity here; we get it into the gut, as Stephen Jay Gould says... But in bringing home the infinite, metaphor also alters the home it's brought to, as Loss Creek is irrevocably altered by my deep time *aporia* ... Metaphor, like the ferry whose etymology it shares ... goes back and forth, to and fro. It renders the infinite tangible, but also infinitizes the here and now.
>
> (129)

I would argue that death's uncanny relationship to aporia and metaphor here (the latter's oscillations resembling Charon's ferrying of souls across the river Styx) is quite significant. Why does McKay suggest that Anthropocene subjects "need" to experience metaphor's power to "bring home" deep time's temporal estrangement? Two key reasons are suggested in the passage above. First, deep time has the power not only to make the infinite "tangible" via metaphor, but also "infinitizes the here and now," underscoring how and why the species-level choices we make today will matter in the future, given they will have considerable temporal duration. However, simultaneously, deep time reminds us about the "gut level" of our mortality, the finite quality of our individual lives. If an all-too-human drive to control or deny death underwrites matérial's destructive appropriations (such as the killing and display of the raven in "Baler Twine"), then deep time's ability to inspire us to confront our mortality is more vital than ever at a time when we are poised to choose between developing increasingly powerful (and potentially destructive) technology in an attempt to "control" the natural world and/or to extend neocolonialist oppression of disenfranchised "others." McKay's geopolitics suggests that accepting the mortality and bodily vulnerability we share with other "members of deep time" might be a step towards developing more ethical socio-ecological praxis. In collections such as *Strike/Slip* and *Paradoxides*, McKay explores deep time's power not only to astonish us, but also to inspire a much-needed alternative to matérial as a basis for home-making.

Readers seeking an illustrative example of how deep time entered McKay's writing "much like the hole down which Alice's rabbit disappeared" (*Angular Unconformity* 558) and what he found on the other side could do worse than exploring "Loss Creek." This poem begins with a series of specific details that evoke the Loss Creek River Fault from the ground up, along with the speaker's paradoxical desire to experience this place "exactly" (directly) in a phenomenological sense—i.e. outside of its meanings in already existing discourse, particularly the post-Romantic tradition of "nature writing"—even though as a poet he must communicate via linguistic mediation:

He went there to have it
exact. The broken prose of the bush roads.
The piles of half-burnt slash. Stumps
high on the valley wall like sconces
on a medieval ruin. To have it tangible.
To carry it as load rather than as mood
or mist. To heft it—earth measure,
rock measure—and feel its raw drag without phrase
for the voice or handle for the hand.

(431)[2]

The depiction of ecological damage here recalls McKay's earlier prose musings on Vancouver Island's Jordan River Main (which is not too far from Loss Creek) in "The Bushtit's Nest." There, the poet bemoans the fact that not much remains of the "Jordan River" except its name, as this body of water was dammed to provide electrical power for the island and the surrounding old-growth forest was clear-cut so severely that the landscape now resembles "a leftover prop from an extremely low budget sci-fi flick about cosmic worms" (*Vis à Vis* 89). Loss Creek likewise bears the scars of clear-cutting, but here the damage is more "tangible" and more resistant to complete erasure under the sign of an "epitaph" than was the Jordan River. I would argue that this is because deep time has shaped McKay's experience in "Loss Creek," transforming his understanding of place in a way that contrasts significantly with his earlier thoughts about the Jordan River basin. In "Loss Creek" the ecological "loss" communicated in the poem's title and opening lines is linked with ruinous human inhabitation not only in the present, but also in a deep historical sense, signaled in the way that the stumps of felled trees resemble "sconces/ on a medieval ruin." The word "sconce" has a double meaning, both as a metal light fixture that supported torches on the walls of medieval buildings, but also as an earthen mound used in military campaigns, suggesting a link between ecological destruction and the history of human *oikeios* as matérial.

Yet the introduction of water in into this landscape seems to afford an alternative way of thinking about temporal flow and human poiesis beyond eco-melancholia at recurring patterns of historical ruin:

He went there to hear the rapids curl around
The big basaltic boulders saying
husserl husserl, saying I'll
do the crying for you, licking the schists
into flat skippable discs. That uninhabited laughter

sluicing the methodically shorn valley.
He went there to finger the strike/slip
fissure between rock and stone between Vivaldi's
waterfall and the wavering note a varied thrush
sets on a shelf of air.

The "voice" of the river rapids seems to waiver between the sound that the water makes when it wraps around basaltic boulders in a phenomenological sense (the play of sound/meaning in "husserl husserl") and a more conventionally anthropomorphic locution aimed at assuaging the speaker's initial feelings of tragic loss at the creek's ecological damage.

What does the "voice" of deep time tell the poet here, brought home by the metaphor of temporal flow that combines water and ancient boulders? From an anthropocentric perspective, the creek seems to say: keep your tears ("I'll do the crying for you"), the world rolls on and should be approached by contemporary humans in a spirit of open-ended play (just as a child might enjoy throwing river stones that have been shaped into "flat skippable discs"). Substituting comedy for tragedy as a means to think about deep time, the speaker can derive comfort at the thought that the wilderness before him is resilient, its "uninhabited laughter/ sluicing the methodically shorn valley," a laughter that opens onto the infinite. But critically, McKay breaks the spell of this all-too-soothing response by drawing attention to the fact that the river's anthropomorphic "voice" is not just a metaphor, but is also the product of ventriloquism—a Romantic projection of the speaker's human desire onto a damaged landscape. The poem's final ten lines inspire doubt about this "voice" by acknowledging that a *rift* or "strike/slip" fault-line exists between human artifice (including poetry) and the radical otherness of more-than-human entities. Mirroring the creek, the speaker touches a basaltic boulder, "finger[ing] the strike/slip," thereby experiencing a "fissure between rock and stone" (a gap between unworked rock and human-altered stone), and "between Vivaldi's / waterfall and the wavering note" of a "varied thrush" (between nature as muzak soundtrack and nature as spontaneous birdsong).

In McKay's oeuvre, the distinction between rock and stone is particularly significant *vis-à-vis* the issue of human dwelling. As he muses in a prose poem titled "Between Rock and Stone: A Geopoetic Alphabet":

> What is the difference between rock and stone? … Many will say there is no difference … [yet] what happens between rock and stone is simply everything human, from the modifications necessary to make homes to, at the other extreme, the excesses of ownership and exploitation which submit all ends to ours. So another answer might be: rock is as old as the earth is; stone is only as old as humanity.
>
> (59)

In the prose poem "Gneiss," McKay contemplates a circle of standing stones at Callanish, Scotland, concluding that our Neolithic ancestors "engineered [a] dangerous dance" when they "required that some of the rocks that comprised their island should stand up with them against the levelling wind and eroding rain"—when, in short, "they insisted that rock be stone." Looking closely at these stones, the poet perceives "the commotion of stress lines swirling within each slab … each stone a pent rage, an agon" (*AU* 453). Archetypically, McKay's writing brings out such easily overlooked fissures, signs of stress that reveal a gap between the more-than-human world's independent ontology and the uses that we humans make of natural phenomena. As Blythe Hutchcroft compellingly demonstrates in "Here at the Brink," McKay is often drawn to "geologic gaps, fissures, and abysses, because [their] missing data 'represents a significant gap—often millions of years' which 'upsets' human authority" (76). The "authority" that is upset goes beyond scientific discourse which desires complete knowledge (an unbroken geological record); rather, it denotes McKay's fascination with ways deep time can constitute a "fissure" more broadly, "through which [it] leaks into history and upsets its authority" (*AU* 9).

What is remarkable about "Loss Creek" is the way that McKay allows deep time to upset his poetic speaker's authority to project a "voice" onto the landscape that releases him from environmental responsibility. The poem's final lines acknowledge that there's a risk in allowing deep time to disrupt the falsehoods we weave with language,

> Recognizing the sweet
> perils rushing in the creek crawling
> through the rock.
> He knew he should not trust such
> pauseless syntax.
> That he should just say no.
> But he went there just the same.

In allowing deep time to create a rift in human hubris, the poet risks not only losing the power of Romantic appropriation, but also any sense that places like "Loss Creek" will conform to our desire to anthropomorphize nature—either as a forgiving surrogate (one who will "do our crying for us") or as a domain that exists for our amusement (as a treasure trove of playthings, like rocks for skipping). Ultimately, though, the poet accepts these risks. In allowing deep time's "pauseless syntax" to disrupt the flow of his language, he gains a new appreciation of "hereness" that is exactly the opposite of Olympian detachment: an intimacy with place that refuses to take flight from the ecological destruction depicted in the opening lines.

Lest readers of poems such as "Loss Creek" be tempted to conclude that McKay is essentially a "place-based" writer with little to say about the Anthropocene's global implications, it is important to examine texts that are not overtly inspired by the poet's immersion in specific environments (such as Vancouver

Island). Indeed, Jesse Patrick Ferguson asserts that McKay's "intense and prolonged imaginative interaction with geology in *Strike/Slip* [the collection where "Loss Creek" first appeared]" reflect a "cosmopolitan" rather than regional sensibility. Ferguson asserts that McKay's geological cosmopolitanism reflects an understanding of globalism's interdependent economic systems and environmental impacts, but also explores a more fundamental "physical model for understanding humanity's interconnectedness" based upon "the spatial conception … of the earth's mantle connecting diverse animals, plants, and human communities, providing a common bedrock from which even the most rarified of our cultural products … literally originate." Moreover, Ferguson also suggests that this sense of a "common bedrock" has significant political implications as an ethos that is "realistic about its limitations and is likely to succeed in mitigating human conflict and encouraging stewardship of the earth" (169).

I would argue that this geological cosmopolitanism is amply displayed in poems such as "Devonian" from *Strike/Slip*. In this text, readers are taken on a temporal as well as spatial odyssey that unfolds on a global stage. The text begins in the intimate space of "theater" that could be almost anywhere, then dilates to plunge readers into deep time, before circling back to the present once more in the final lines. The phrase "four hundred million years" casually uttered by someone at a theater performance initiates a strange journey whereby

> the words
> tap dance with their canes and boaters through
> that spotlight right across the stage unspooling out the
> stage door down the alley through the dark
> depopulated avenues (for everyone is at the theater) toward
> the outskirts where our backyards bleed off into
> motel
> rentall
> stripmall
> U-haul past willowscruff, past ancient
> -rusting mercuries along the lovers' lanes the coyotes lure
> our family pets down all those creekbanks …
> […]
> … the dried-up
> memories of water how the waves were how
> the light that fell so softly through the depths was
> intercepted by the lobe-finned fishes flickering among the
> members of the audience still staring at the empty stage—four
> hundred *million* years, yes, that's a long
> long time ago.
>
> (*AU* 438)

Deep time's disruption in this text is less place-specific and its effects (on the surface) are less immediately relevant to environmental politics than was the case in "Loss Creek." Yet despite the slapstick quality of the reeling mind's exit from the bourgeois "theater" of human affairs when it is confronted with "Devonian" deep time, it's suggestive that this comic flight is also an odyssey through what might be characterized as an uncanny Anthropocene "wilderness." Once the speaker's mind leaves the brightly lit stage (of affluent entertainment culture), he is propelled into "dark / depopulated avenues," the desolate "outskirts" beyond the comfortable "backyards" of global capitalism's economic elite.

This tour of the bourgeois collective unconscious is visually excavated on the page like layers of geological strata. We visit an anonymous "motel," followed by other sites that suggest rootless mobility (rental outlets, a strip mall, a U-haul store) which repeat the same "al" sound in descending order, until we come to rest in a wilderness of "rusting mercuries" (end of the line for car-culture) and resilient animals (like coyotes) capable of surviving in dry, climate-change-altered desertscapes. Readers are invited to contemplate a future wherein present-day decomposing autos have become techno-fossils, perhaps uncovered by future geologists as artifacts of petromodernity's ruins. Having surveyed these signs of contemporary catastrophe, the final lines conjure ghosts from the Earth's distant past, returning us to the Devonian period some "four hundred million years ago" as "lobe-finned fishes flicker among the / members of the audience." Is this surreal merger of deep time and the present moment a reminder of the shared evolutionary origin of human beings and fish (common "memories of water") or does it suggest that, before long, our species might be joining "lobe-finned fishes" in the roster of the extinct? After all, the late Devonian was marked by a series of die-offs that, cumulatively, marked one of the planet's five largest extinction events. In the midst of the Anthropocene's sixth great extinction "members of the audience" (who double as readers of the poem) are left "staring" at a stage that appears "empty," perhaps because its star player, *Homo sapiens*, has exited. Yet the vacant stage also raises questions about whether its "emptiness" reflects a problematic anthropocentrism on the part of the audience—an arrogance that would see an Earth without our species as null and void, rather than acknowledging the endurance of other lifeforms, other actors with whom we share the stage of existence.

Indeed, a survey of McKay's work suggests that one "gift" that has come out of his contemplation of deep time's dangerous de-familiarization has been the challenge this has posed to what might otherwise have become a complacent "comic anthropocentrism" in his earlier work. Part of becoming "members of deep time, along with trilobites and Ediacaran organisms" requires not only giving up "our special status as Master Species" but also learning to see ourselves as only "one expression of the ever-evolving planet" ("Ediacaran and Anthropocene" 53). McKay diverges from Jeffers, however, in suggesting that giving up anthropocentric "mastery" to acknowledge our membership in deep time doesn't mean that what we do as a species in the present historical moment will be negligible in the long run. Indeed, in poems like "Loss Creek" and "Devonian," deep time heightens attention to present-day destruction, asking us to contemplate the choices we are making today about the Anthropocene's socio-ecological legacy, what kind of world we are leaving to future generations (human and more-than-human alike). Thus, the empty stage that concludes "Devonian" seems to pose the question: will Anthropocene subjects be "poor players" who "strut and fret [their] hour upon the stage," but who ultimately leave "nothing" behind but a legacy of nihilistic destruction?[3] As a species, we are capable of understanding the trace of past geological catastrophes recorded in fossils; but as "members of deep time" we are also uniquely capable of acting with self-awareness and ethical choice concerning whether our actions will (collectively) constitute another far-reaching global catastrophe. In creating texts that reflect upon this dilemma, contemporary geopoets such as McKay allow ancient stone to "speak" to us in ways that have never been more timely. Hence, rock's strange portent in poems such as "Mistaken Point" from *Paradoxides.* Contemplating the fossilized remains of Precambrian lifeforms that existed 560 million years ago before they were wiped out by global volcanic eruptions, the speaker unexpectedly finds something relevant to us, "[b]ack here, / in the Anthropocene," something that may also be a sign of future things to come. He provocatively asks

> What shall we call
> this antique frond, part fern, part feather,
> part Art Nouveau and brand new Braille,
> urgent and enigmatic as an oracle?
> (519)

Angular Unconformity II: Dionne Brand and the Problem of Anthropocene Universalism

Thus far in this book, I have examined divergent interpretations of geological deep time's political significance in Holocene versus Anthropocene-era aesthetics. From a scientific standpoint, it is not necessarily problematic to study our species-level evolution and planetary impacts within the context of geological time, yet it is precisely such a focus on species-level agency that has prompted significant backlash among humanities scholars who take issue with Anthropocene theory's universalist constructions of political agency. A very diverse configuration of thinkers, from Andreas Malm and Alf Hornborg to Anna Tsing, Jason Moore and Donna Haraway have registered important objections to Anthropocene theory that construes "Anthropos" (the human species) as an undifferentiated mass that is equally responsible for wreaking geological-level damage on the planet. Such constructions of geological agency tend to ignore the outsized role that historical and present power structures—particularly those associated with capitalism, colonialism and petromodernity—have played in bringing about the Anthropocene's environmental and social crises. A more comprehensive engagement with the politics of geological deep time would take into account how Holocene-era practices such as slavery, colonialism, environmental racism and fossil-fuel energy regimes have shaped our present historical moment, and will continue to do so long into the future unless we collectively intervene to stop this toxic socio-ecological legacy. Kathryn Yusoff's ground-breaking postcolonial deconstruction of the "White Geology of the Anthropocene" in *A Billion Black Anthropocenes or None* represents such an intervention, given its trenchant examination of ways that colonialist geology has constructed black fossil-fuel reserves and the bodies of people of color as "inert," extractable resources. Indeed, if thinkers such as Chakrabarty are correct in asserting that one of Anthropocene theory's primary challenges is to bring "anthropos" (species-level) agency into a more productive dialectic with "homo" (historical-level) politics, then studies like Yusoff's—that reveal toxic interfaces between geological materials produced over the course of deep time (such as coal and gas) and the more recent (and ongoing) history of social oppression associated with energy regimes that extract such materials—is a much-needed corrective to potential white-washing entailed in universalist declarations that the only way forward is for geopolitics to assume that "we're all in this together" in the climate-change era.

Significantly, both Jeffers and McKay focus primarily on "anthropos"-level understandings of deep time, and neither hesitates to employ a universal "we" in making declarative proclamations about things such as human depravity (Jeffers) or human "membership" in deep time (McKay). Arguably, the status of both poets as white, economically privileged men who are descended from settler colonialists has influenced the invisibility of identity politics in their poetic musings about deep time as both occupy a subject position that has historically been assumed to have the authority to make such universal proclamations. But how might a female, Marxist, queer poet of Afro-Caribbean descent such as Dionne Brand understand the politics of geological time *otherwise*—as neither implying equal historical culpability nor automatic "membership" in the community of beings granted ethical treatment? Although Brand shares Jeffers's fascination with sea and stone as elements that inspire contemplation of deep time, in texts such as *The Blue Clerk* she also critiques the idea that poets should avoid engaging with history or politics when addressing "nature" as a sign of universal truth. As the poem's acerbic "clerk" notes in her ongoing dialectic with the "author," the latter's tendency to merely "inventory" human suffering or to seek aesthetic escape from historical trauma by focusing on timeless "matters of nature" actually reflects an all-too-familiar, jaded Romanticism. As the clerk suggests, no historical person (not even poets!) can actually occupy the subject position of the eternal, any more than individuals can actually live their lives hovering above the Earth from a position of Archimedean objectivity. Likewise, although contemplating geological time's vast temporal expanses can have the effect of "decentering" anthropocentric hubris in ways that both Jeffers and McKay celebrate, Brand's poetry reminds readers that deconstructing anthropos-level agency alone is politically insufficient and must be supplemented by a "homo"-level understanding of historical forces that have perpetuated environmental and social injustice. By exploring links between extractive industries, (neo)colonialist oppression, ongoing racial injustice and the relationship of all of these things to climate change, Brand's poetic explorations of geological time underscore the vital importance, rather than the negligibility, of present political policy. As such, her verse constitutes what Yusoff calls an "insurgent geology of belonging" (88)—a "poetry of rocks that tells a different story of rocks" (93) than the ones told in Jeffers's and McKay's lithic lyrics. In the remainder of this chapter, I will explore two ways in which Brand's verse reflects this "insurgency": namely, her excavation of ways geological deep time meshes with colonialist histories of slavery/extraction and her deconstruction of the geological sublime as a psychic "escape" from history. Although Brand's poetry often rehearses the appeal of inhuman time and space as a response to historical trauma, she is acutely aware of the inadequacy of such impulses, ways they can impede the urgent need to formulate a truly new geopolitics for the Anthropocene era.

If sea and stone are natural elements often evoked with great sensory vitality in Brand's writing, industrial mineral extraction and oil production constitute an equally prevalent, though uncanny, presence in her texts. It is tempting to speculate about how the author's birthplace in Guayaguayare, Trinidad may have influenced this, given this village's history as a major site of petroleum development from 1902 onwards, when the first commercial oil wells were drilled in the area. This industry continues in the present as companies such as BP Trinidad & Tobago control the harbor at Guaya, making it a regional hub for offshore oil production. Although it is the sea (rather than earth minerals or petroleum) that is immediately foregrounded in Brand's memoir *A Map to the Door of No Return* as "the first thing" (8) in the speaker's childhood memories, its status as a sign that transcends history is ultimately deconstructed in the text, just as geological time is in other poems. As a child, the speaker believes "the sea was its own country, its own sovereignty" (7) apart from humanity, and feels

> it would forever be larger than [her] … [reducing] all life to its unimportant random meaning. Only we [villagers] were changing and struggling, living as if everything was urgent, feeling—the ocean was bigger than feeling … It took our happiness as minor and transitory.
>
> (12)

Yet, she also senses that this same 'inhuman' force is also implicated in her people's history, stating, "Our origins seemed to be in the sea. It had brought the whole of Guayaguayare there from unknown places, unknown origins" (12). Though from a deep-time perspective the ocean is the "origin" of all life on the Earth, the speaker comes to realize that the sea has also been a conduit for the more recent history of slavery, French colonists having imported Africans (including her own ancestors) to work as slaves on plantations in Trinidad & Tobago. From this perspective, the sea is less "its own country" than a source of traumatic dis-placement, an element that helped to erase the speaker's African origins in the service of a monstrous trade that leaves "wounds" in Guaya's black residents that have not healed over time:

> I knew that everyone here was unhappy and haunted in some way… I had a visceral understanding of a wound much deeper than the physical, a wound which somehow erupted in profound self-disappointment, self-hatred, and disaffection.
>
> (11)

Although, as in Jeffers's poetry, the speaker in Brand's memoir is attracted to the sea's sublime indifference to human suffering as a psychic escape from trauma [noting its ability to "wash away blood and heal wounds" (10)], she cannot ultimately escape the realization that, in Walcott's words, "the sea is history," admitting she "knew that before [she] knew it was history [she] was looking at" (12).

As a poet writing in the romantic lyrical tradition but also one who significantly subverts this tradition, it is important to note how Brand repeats this pattern of evoking/undercutting the inhuman as a source of solace in the face of socio-political injustice throughout her oeuvre. This pattern also applies to her evocations of deep time and the geological sublime. For all the sophistication of Yusoff's interpretation of Brand's "insurgent" verse, the poet's thoughts about deep time remain largely unexamined in *A Billion Black Anthropocenes or None.* Yusoff's interest in how Brand's verse "cuts through coloniality as a counteraesthetic that refuses the inhuman in its codified state as property" (88) leads her to focus almost exclusively on how the poet challenges the reification of black bodies and black minerals as "resources" for colonial exploitation. Yet, I would also argue that Brand's critique of relatively recent colonialist history is accompanied by an equally trenchant critique of the romanticization of geological deep time as something that escapes human history, or is (of necessity) separate from politics.

In *Land to Light On*, Brand depicts the wanderings of a diasporic subject who feels alienated due to the misogyny, racism and homophobia she encounters in Canada as a female, black and queer immigrant. Postcolonial interpretations of this poem tend to focus on what Katherine McKittrick describes as "the predicament of black geographies as they are understood beyond and through colonial and imperial projects" (105). McKittrick interprets the speaker's attempts to "map" Canada in the poem as "chart[ing] a different sense of place, which is simultaneously unexpected, rooted and rootless … because a [diasporic] black claim to place is unavailable through traditional geographic means" (104). Disoriented in a cold white landscape where she "see[s] nothing recognisable" (4), the speaker is assaulted by obscenities from passersby, such as "a white man in a red truck" who "scream[s] his exact hatred … something about [her] cunt" and, on another occasion, harassment from a white policeman who stops her and two other people of color without cause, regarding them as "aliens" (73). In "Dionne Brand's Environmental Poetics" Cheryl Lousley argues that the poem's "inscription of blackness in [Canada's] rural areas and wilderness is provocative, since non-urban landscapes tend to function metonymically to represent the territory of the nation as a whole" (47) and wilderness archetypically represents a flight from history—particularly the history of European cultures' systematic erasure of indigenous peoples from "wilderness" areas.

Yet Brand's verse is also provocative in bringing to the reader's attention another, equally ubiquitous, source of alienation in her cross-country drives: namely, petromodernity and its connection to ongoing neocolonialist exploitation. It is no accident that many of the speaker's journeys take place by car, along lonely country roads in the dead of winter. Early on in *Land to Light On*, the speaker describes the vertigo she experiences one afternoon while "driving north / on 35, stopping for a paper and a coffee" (13) at a gas station. Instead of a relaxing break from a long journey, the speaker is stunned by "the terrifying poetry" of newspaper headlines:

> the wave of takeovers, mergers and restructuring
> ... swept the world's ... blue chips rally in New York
> ... Bundesbank looms ... Imperial Oil increases dividends
> ... tough cutbacks build confidence

These headlines trigger an experience that is at once profoundly disorienting and oddly clarifying:

> You feel your coffee turn asphalt, you look around
> and your eyes hit the dirty corners of the windy store
> stray paper, stray cups, stray oil, stray fumes of gas.
> Your mouth never opens, your keys look unfamiliar.

Brand de-familiarizes the ostensibly mundane experience of a gas station rest-stop by setting in motion a complex series of associations that expose petromodernity's larger networks and their hidden socio-ecological costs. Significantly, both oil and coffee production have roots in colonialism, and both industries are associated with exploitative labor practices that disproportionately affect people of color and produce environmental destruction in developing nations. As Rob Nixon's *Slow Violence and the Environmentalism of the Poor* documents, multinational petroleum companies such as Shell BP have a history of exploiting local workers in countries such as Nigeria (paying workers less than $1 per day as of 2011), bribing police to quell labor uprisings, and polluting the local water, air and land such that traditional subsistence farming and fishing becomes impossible. Likewise, uncertified coffee production in developing nations often entails unsustainable harvesting practices (such as the clearing of large areas of rainforest which reduces biodiversity and contributes to climate change) and taking advantage of migrant labor in an attempt to mitigate increasingly unpredictable crop cycles. In lines such as "[y]ou feel your coffee turn asphalt," Brand makes visible the ways that coffee consumption in wealthy nations such as Canada is tied to fossil-fuel-powered transportation networks that carry beans across great distances. Likewise, her speaker is struck by signs of pollution in the station—"stray paper, stray cups, stray oil, stray fumes of gas" (15)—that are uncanny, unwanted reminders of petromodernity's environmental costs, eclipsed by newspaper headlines celebrating the "increasing dividends" for companies like "Imperial Oil." Such heightened awareness makes the speaker's car keys "look [suddenly] unfamiliar," inspiring her (and the poem's readers) to question the extent to which consumers in developed nations who routinely enjoy the caffeine-fueled "freedom" rides actually participate in wider systems of oppression.[4]

Other sections in *Land to Light On* delve deeper into the past in order to clarify connections between racism, extractive industries, environmental destruction and colonialism's afterlives. In section XIII, a black speaker's all-too-common experience of being stopped by a white cop for no reason other than racial profiling triggers thoughts about historical forces that combine to produce such encounters:

> In this country where islands vanish, bodies submerge,
> the heart of darkness is these white roads, snow
> at our throats, and at the windshield a thick white cop
> in a blue steel windbreaker peering into our car, suspiciously
> … or perhaps
> not suspicion but as a man looking at aliens.
> Three Blacks in a car on a road blowing eighty miles an hour
> in the wind between a gas station and Chatham. We stumble
> on our antiquity. The snow-blue laser of a cop's eyes fixes us
> in this unbearable archaeology.
>
> (73)

Stopped en route between another "gas station" and "Chatham," the speaker and her travelling companions are forced to confront law enforcement's institutional racism as personified in a white cop's penetrating, "laser"-like stare. The reference to "Chatham" suggests that the policeman's racial profiling is symptomatic of slavery's afterlives, given the history of the Chatham-Kent municipality of Ontario as a community established by African American slaves who escaped to Canada from the US, and of Chatham Manor (located in Virginia) as a major plantation and site of a famous slave uprising in the winter of 1805. Thus, though the black speaker and her friends "stumble on [their] antiquity" when confronted by police harassment as a symptom of ongoing social injustice, she, like prior generations of blacks who rebelled against slavery, refuses to be reified by this "unbearable archaeology." Instead, Brand will conduct her own "archaeology" which will bring to the surface "submerged" connections between racial injustice, extractive energy regimes (signified by the "gas station") and "the heart of darkness [that] is these white roads," routes which (for better or worse) her party must hazard.

The evocation (and inversion) of Manichean racism in Conrad's *Heart of Darkness* in line 2 is particularly effective as a means to illuminate colonialism's deleterious effects upon people of color and upon the environment. Revising imperialist attempts to project the "heart" (or essence) of "darkness" onto Africa's wilderness or peoples, Brand's speaker excavates the traumatic history of colonialism's "darkness" in Canada:

> That cop's face has it. "They had been in this vast and dark
> country only a short time." Something there, written as
> wilderness, wood, nickel, water, coal, rock, prairie, erased
> as Athabasca, Algonquin, Salish, Inuit … hooded in Buxton
> fugitive, Preston Black Loyalist, railroaded to gold mountain,
> swimming in *Komagata Maru* … Are we still moving?
> Each body submerged in its awful history. When will we arrive?
>
> (77)

Like the earlier reference to *Heart of Darkness*, the quote from Conrad's "An Outpost of Progress" cited in this stanza illuminates ways that European colonialism racialized concepts of wilderness and dark-skinned peoples as "others" to white "progress" in order to justify resource exploitation, slavery and genocide. As Lousley observes, the poem "exposes wilderness as a claiming of territory for the sake of resource exploitation, which, like Aboriginal dispossession, is hidden under the cultural belief in wilderness as untouched nature" ("Environmental Poetics" 48).

Though it is certainly the case that Brand casts a broad net in the passage above to bring the twinned history of natural resource "development" and racism in Canada to the surface, the poem as a whole is particularly invested in excavating the history of colonialist *geological* extraction and black slavery, along with the legacy of these practices in contemporary petromodernity. Thus, the "nickel," "coal," "rock" and "gold" signifying extractive industries in this passage and the references to slavery ("hooded Buxton fugitive, Preston Black Loyalist") are reiterated in other sections of the poem, reaching a climax in section XV's meditation on "Every Chapter of the World." There, the speaker concludes that she can "never measure the length of this / tongue of conquest" used as she is "to the mind's / enervating sleep on this whole chain of centuries":

> she knows every chapter of the world describes
> a woman draped in black and blood, in white
> and powder, a woman crippled in dancing and
> draped in dictator's dreams, in derelicts' hearts,
> in miners' lights, in singers' shoes, in statues,
> in all nouns' masculinities, in rocks cut out in
> every single jungle and desert secret carried
> (95)

A sense of futility in cataloguing wrongs repeated throughout history emerges in this final section, culminating in the statement "Inventory is useless now" (101). Yet, in exposing cumulative strata of injustice, the poet illuminates patterns which might otherwise remain hidden just as "miners' lights" reveal seams in the earth which record past cataclysms.

Given the pain entailed in Brand's poetic archaeologies and her frequent frustration with how efficacious these excavations are in a world that seems to endlessly reiterate socio-ecological injustice, there has been considerable debate about the political import of her work. While critics such as George Elliott Clarke have taken Brand to task for indulging in a "brokerage poetic" (41) whose didacticism never transcends Romanticism, others such as Sophia Forster suggest that Brand's political "ambivalence" reflects her ability to perceive "all too well not only the strengths but the political weaknesses of two of language's major (and opposing) capabilities, the creation and subversion of identity structures" ("Politics of Ambivalence"). Ecocritical readings of Brand's poetry tend to be less equivocal in asserting the political importance of her work. For example, Catriona Sandilands's "Acts of Nature: Literature, Excess, and Environmental Politics" lauds Brand's

Land to Light On as a text that "insists, in both content and form, on a political reading" (139)—one that represents

> the north as a landscape of ongoing violence, as a place that is not a refuge from but rather profoundly a part of a global web of ongoing violences that connect to one another through capital flows and individual stories of coming and going.
>
> (140)

Likewise, Christine Okoth's "The Extractive Form in Contemporary Black Writing" expands upon Yusoff's work to interpret Brand's *Inventory* as a text that reflects literature's ability to not merely "play a … diagnostic function, pointing to sites of extraction across the globe and lamenting the state of ecological damage" but rather, critically, to "identify the forms that extraction produces and [to] appropriate those same forms in pursuit of both political and literary formal reinvention" (391). While I share these critics' conviction that Brand's insights about social and environmental injustice are critically important, I would also suggest that one reason the political import of her work has prompted such controversy is the subtlety with which she stages a dialectic of historical trauma and a desire to psychically escape from such trauma via recourse to sublime, elemental earth forces such as the sea, weather and rock. Geological deep time constitutes one of the most significant of these forces given its relationship with colonialist history and extractive energy regimes.

Given Brand's trenchant critique of this history in *Land to Light On*, passages in which the lyrical "I" seeks spiritual solace via recourse to inhuman geology are surprising and unsettling. For example, in section IV (xiii), a speaker who feels "heavy" in her "inanimate body, all stone" muses

> I was always for succumbing to something bigger,
> I wanted something unindividual like distance or chaff
> … and undifferent as obsidian caves bluing and blacking
>
> (39)

Significantly, the speaker goes on to say that she desires these things because she "had hoped," that through them, "it would occur to [her] / that nothing was wrong." Within the larger context of historical and present-day trauma that mobilizes the poem as a whole, one could read such passages as reflecting a longing for quiescence, a release from pain via an imagined merger with the stone and its "bigger" inhuman temporal frames. Wanting "undifferent" things suggests a desire for emotional ossification, a release from intolerable affect made possible by adopting an "undifferentiated" attitude of apathy or indifference. More radically, the speaker's desire here to "succumb" to "something unindividual" could also be read as a desire for death as the ultimate release from violent temporal existence and/or from oppressive politics which denigrates difference. Hence one could interpret the speaker's attraction to "undifferent obsidian caves bluing and blacking" as reflecting a desire for death-like peace (the cave symbolizing a grave) in order to escape suffering (being bruised "black" and "blue" by life). Yet I would argue that these "undifferent" obsidian caves can also be read as reflecting the allure of the geological sublime in that their dark stone impartially records catastrophic events (volcanic eruptions) that have been displaced to the distant past instead of endured in the immediate present. As such, they might give the speaker enough "distance" on more recent historical trauma to conclude that, in the larger scale of things, "nothing" is ultimately "wrong" because nothing is permanent from a deep-time perspective.

However, what's truly remarkable about the obsidian caves in this stanza, like many other complex tropes in Brand's verse, is that they also *deconstruct* the very longing they instantiate. Given Brand's interest in the Black Diaspora and in African history, it is likely that the "obsidian caves" are a reference to the Paradise Lost Caves near Nairobi, Kenya. Discovered by Joseph Mbai on his property in 1996 (the year before *Land to Light On* was published), these 2.5-million-year-old caves contained human remains dating back 8,000–12,000 years and many obsidian artifacts (including hunting implements) from the Later Stone Age. Moreover, although the caves could be seen as a manifestation of geological time and of Africa's deep history, they also are also important in its more recent colonialist history. During the 1940s, members of the Mau Mau Kenyan liberation movement used the caves to plan actions against the British Colonialist government and would hide/regroup in them during the 1950s after clashes with British forces in Nairobi. So, although Brand's "obsidian caves" may initially appear to evoke the geological sublime as an *escape* from history, from another perspective it would be more accurate to say that they can be seen as a *locus* of history, gathering together geological time, Africa's pre-historic past and its more recent struggle for independence from colonialist rule. Indeed, the Mbai caves can be read as a material embodiment of the merger of earth time and historical time which is a key challenge in contemporary geopolitics. Perhaps this is why they feature so prominently in the work of visionary Kenyan architects associated with Cave_Bureau, including their ambitious *Anthropocene Project*. In the words of Kabage Karanja, one of the group's co-founders, these architects are

> trying to reflect on that geological term, the Anthropocene, the new age of man, trying to look … at the colonial era, the history of this age that we live in that has been the defining factor of where we are as a civilization. And we feel having another sort of layer and a voice over that narrative is actually very critical.
>
> (quoted in Brownell)

The same could very well be said of Brand's critical project in excavating links between geology and colonialist history in *Land to Light On* and other more recent long poems such as *Inventory* and *Ossuaries*.

In *Inventory*, Brand expands her focus on New World colonialism in *Land to Light On* in order to encompass a larger, global perspective. Initially a response to

US wars in Iraq and Afghanistan and to attempts to document civilian casualties which the US military refused to track (such as the Iraq Body Count Project), the poem as a whole is a deliberately depersonalized attempt to catalogue (or take stock of) a broad spectrum of social and environmental injustice at the turn of the twenty-first century. As Lousley's "Witness to the Body Count" documents, "*Inventory* extends the scope of its ethical project beyond human war and suffering to the ecological realm with the metaphorical inclusion of the planet itself in the 'body count'" (46). The poem documents the numbing effect of habitually witnessing televised violence from a safe distance, from 9/11 to wars over oil to the aftermath of Hurricane Katrina. Yet, as in Brand's depiction of her gas station epiphany in *Land to Light On*, she also subverts a sense that responsibility for socio-ecological disaster always lies 'elsewhere' by depicting the complicity of North American cities in petromodernity and violent conflicts over fossil fuel:

> Observed over Miami, the city, an orange slick blister,
> the houses, stiff-haired organisms clamped to the earth,
> engorged with oil and wheat,
> rubber and metals …
>
> (15)

Parasitic in their reliance upon "oil" and mineral extraction ("metals"), such cities are emblematic of why many large urban areas in the United States are at once ecologically unsustainable and reliant upon foreign wars to maintain artificially low energy prices and uncapped consumption.

Brand's critique of such complicity is further radicalized in section III's depiction of the speaker's own cybernetic freeway experiences on the day 9/11 occurs:

> It was not coincidence then that the day
> was beautiful, the highways roaring …
> machine and body, shield and tissue,
> the highway worked itself into her shoulders …
> now she was concrete and car, asphalt and oil,
> head whirring like any engine,
> becoming what they were becoming …
>
> (45–6)

Remarkably, the speaker depicts her own complicity with petromodernity in such passages, as the resident of large cities such as Toronto and as a poet who frequently writes about her car journeys. Given the overwhelming trauma of events such as 9/11, the loss of life that occurred on that day and afterwards in Middle Eastern wars and terrorist bombings catalogued with excruciating detail in *Inventory*, it is not surprising that at certain moments the speaker turns away from this violence to contemplate the more-than-human world's relative sanity:

the physical world is not interested in us,
it does what it does,
its own inventory of time, of light and dark …
inexplicable then our certainty
in the fame of politicians and the blackmail of priests
memory will tell us this was foolish,
only, we're not there …

[…]
that ravaged world is here
and the day is always beautiful somewhere …
(46–8)

Once again, these lines brilliantly evoke and undercut the appeal of deep time as a flight from history. The ostensible comfort the speaker derives from contemplating that "the physical world" (the material earth) is "not interested" in our species, but instead has its own, infinitely larger, "inventory of time" is right out of Jeffers's playbook, particularly given the placement of these lines as an interruption of the speaker's meditation on 9/11 and its aftermath. Yet, although from a deep-time perspective humanity's desire to celebrate politicians and to follow priests may appear "foolish," the speaker knows that people can't live their lives in geological time's "elsewhere"—indeed, she states emphatically that "we're not there" occupying a space of Olympian detachment, but rather "here" in an increasingly "ravaged world" that demands our engagement. Indeed, *Inventory* as a whole suggests that the risk of replicating "all that noise that poets make about / time and timelessness" (33) is that the need to remain present and politically active might be lost at a juncture when such engagement is more important than ever. Hence, the poet chooses to remain open to suffering and renounces escapist geo-sublimity in the poem's concluding lines:

there are atomic openings in my chest
to hold the wounded,
besides the earth's own
coiled velocities, its meteoric elegance,
and the year still not ended,
I have nothing soothing to tell you,
that's not my job,
my job is to revise and revise this bristling list,
hourly
(100)

Although contemplating the Earth's powerful cyclical rhythms (its "coiled velocities") and sublime beauty ("meteoric elegance") may be restorative, the speaker maintains it is not the poet's "job" to "sooth" her audience into a state of political complacency by celebrating such forces. Instead, *Inventory* suggests that spiritual hardening is a major occupational hazard for poets who strive to remain politically engaged in the Anthropocene era.

Brand's ability to skillfully represent this dilemma reaches a powerful zenith in *Ossuaries.* This long poem narrates the flaneur-like wanderings of Yasmine, an alienated former Marxist who records her international travels, love affairs, involvement in a bank heist, and her ultimate employment in a US poultry-processing plant in temporally fragmented, often chaotic stanzas. As such, the poem is an infernal odyssey through what Anne Quéma (following Achille Mbembe) characterizes as geoscapes shaped by "necropower"—"death-worlds, new and unique forms of social existence in which vast populations are subjected to conditions of life conferring upon them the status of the living dead" (Mbembe quoted in Quéma). Quéma argues that "it is precisely because Brand has an intimate knowledge of colonial necropower that she is in a position to write a sharp critique of genocidal violence that has become planetary" and notes how the collection's 25 sections function as ironic lyrics, "songs of necropolitics poised between carnage and commemoration." To this analysis I would add that, of all Brand's longer poems, *Ossuaries* seems most explicitly *shaped by stone*—in its depiction of Yasmine's spiritual ossification, its excavation of how "colonial necropower" shapes her life (and the lives of other persons of color living under neoliberal capitalism) and in the shadow of anthropos's future extinction that is cast over the text as a whole.

Indeed, as the poem's title suggests, each section functions as "container" in which the remains of the dead are memorialized, whether this death is spiritual or material (or both). In Ossuary I, Yasmine laments that though she "lived and loved, some might say / in momentous times," in hindsight her "dreams were full of prisons" (9):

> I had nights of insentient adjectives,
> shale nights, pebbled nights, stone nights,
> igneous nights, of these nights, the speechlessness
>   I recall, the right ribs of the lit moon,
> the left hip of the lit moon,
> what is your name they asked, I said nothing
> I heard the conspiratorial water,
> I heard only stone, I ate her shoulder,
> I could not hear myself, you are mistaken I said to no one.
> (14)

Yasmine experiences herself as "insentient" and "speechless" as stone, unable to "hear" herself or even to respond to a skeletal moon's "questions" (which are really her own apostrophic queries posed to a dead celestial rock). This depiction of acute paralysis is immediately followed by lines that suggest the speaker's spiritual deadness is related to the history of extractive colonialism:

after consideration you will discover, as I,
that verbs are a tragedy, a bleeding cliffside, explosions,
I'm better off without …
… such cloth this compass
of mine, of earth, of mourners of these
reasons, of which fairgrounds, of which theories
of plurals, of specimens of least and most, and most
    of expeditions,
then travels and wonders then journeys,
the photographs and photographs of course
the multiplications of which, the enormity of this,
and drill-bits and hammers and again handcuffs,
and again rope, coarse business …

(14–15)

In Yasmine's stream of consciousness, propaganda celebrating colonialism as an engine of "progress" bleeds into imagery reflecting its horrific social and ecological destruction. It is as if imperialism's super-ego projections (enlightened, "compass"-bearing Europeans documenting their "wonderous" expeditions with "photographs") are at war with the truth of its historical Id (the "cliffside explosions," "drill-bits" and "hammers" of mining operations; the "handcuffs," "rope" and the "blood" of enslaved workers), each of these voices struggling within Yasmine's psyche as she tries to "hear" herself and to break through her petrification.

Significantly, unlike Brand's earlier long poems, there is no moment in *Ossuaries* when the narrator finds spiritual relief in imagining a flight into inhuman, geological time. Instead, a melancholy awareness of historical impermanence and of extinction as a possible future for the human race hangs like a pall over the poem. In Ossuary XII, Yasmine thinks about the ruins of Utica, Tunisia as she drives her car to Utica, New York, contemplating how the former "doesn't exist anymore," having been "deluged in a thousand years of silt" (90) and seeing a parallel fate in store for the latter: "she knows what no one here will know / how in four millennia only one wall or two will survive" (92). Though in Ossuary XIV the speaker finds "brief relief" in her realization that "the only thing that amazes her now is the earth, / its ubiquitous snows and lights, / and waters, its combustible air … its miraculous colours" (116–17), the momentary vitality she feels here is almost immediately undercut by the anti-pastoral landscape that surrounds the abattoir where she works:

take the train toward the river's bridge,
toward the electric fields the diagonal bungalows,
graveyards of car tires, and still vines, marsh
keeps struggling up, waste like costal cartilage,
… she's tired of struggling up

(117)

Though nature (the marshland) and Yasmine herself show remarkable resilience in continuing to "struggle up" through petromodernity's wastelands (signified by car tire graveyards), there's also a palpable sense of exhaustion with this effort, such that by the final ossuary (XV) the speaker seems resigned not only to her own petrified existence but also to the imagined end of *Homo sapiens*:

> they ask sometimes, who could have lived,
> each day,
> who could have lived each day knowing
> some massacre was underway, some repression,
> why, anyone, anyone could live this way,
> I do, I do
>   anyone, I'm not unique …
>   so here we lie in our bare arms,
> here the ribs for a good basket, a cage,
>   the imperishable mandible, the rhetorical metatarsals …
>   here we lie in folds, collected stones
> in the museum of spectacles,
> our limbs displayed, fract and soluble
>
> (122–4)

In bearing witness to how an "I" living in late capitalist petromodernity can become habituated to "massacres" and "repression" as the stuff of "everyday" life, Yasmine attests to the fact that she is "not unique." Yet, the poem as a whole suggests her emotional ossification is the tangible result of colonialism's ongoing socio-ecological oppression, a legacy that can produce stifling political petrification and, if unaltered, will likely contribute to the extinction of humanity as a whole.

Hence, the final lines imagine a collection of human bones viewed from a future perspective when *Homo sapiens* is nothing but a relic "in the museum of spectacles," images which uncannily echo those in Ossuary I when the speaker imagines herself "fall[ing] dead like a specimen, / at the anthropometric spectacles / on the Champ de Mars, the Jardin d'Acclimatation" (13). As a person of color, she must routinely negotiate the specter of racist colonialism, symbolized in sites such as Paris's Jardin de l'Acclimatation Anthropologique where, between 1877 and 1912, Africans were exhibited in "human zoos" as they "acclimatized" to life in Europe and were subjected to pseudo-scientific morphological studies. Yet, such grotesque spectacles also forecast a doomed future for all those negotiating the "cage" of ongoing neocolonialism, petromodernity, climate disruption and global necro-politics. It is against the inevitability of such a future archeology that Brand stages her insurgent geopolitics, converting stone into bone, hands into incandescent, revolutionary art:

> were this a painting, it would combust canvases,
>   this lunate pebble, this splintered phalanx,
> I can hardly hold their sincere explosions
>
> (124)

Notes

1 UN DESA. *World Population Prospects 2024: Summary of Results.* July 2024. https://un.org/development/desa/pd/world-population-prospects-2024.
2 All McKay poems cited are from re-printed versions in the anthology *Angular Unconformity* (*AU*), unless otherwise noted. Because these poems do not have line numbers, I give page references from this anthology rather than line citations.
3 *Macbeth* Act V, sc. 5.
4 For a trenchant study of links between the global coffee trade and the exploitation of migrants, see Joseph Nevins's article "Dying for a Cup of Coffee? Migrant Deaths in the US–Mexico Border Region in a Neoliberal Age."

References

Borges, Jorge Luis. Avatars of the Tortoise. In *Labyrinths*, New Directions Publishing, 1964.

Brand, Dionne. *The Blue Clerk.* Duke University Press, 2018.

Brand, Dionne. *Inventory.* McClelland & Stewart, 2006.

Brand, Dionne. *Land to Light On.* McClelland & Stewart, 1997.

Brand, Dionne. *A Map to the Door of No Return: Notes to Belonging.* Doubleday, 2001.

Brand, Dionne. *Ossuaries.* McClelland & Stewart, 2010.

Brownell, Ginanne. An Early Fascination with Caves Leads to a World Stage. *New York Times*, May 21, 2021, www.nytimes.com/2021/05/21/arts/cave-bureau-kenya-architect-venice-biennale.html.

Clarke, George Elliott. Review of *Land to Light On* by Dionne Brand. *Journal of Canadian Poetry*, Vol. 14, 1999, pp. 37–42.

Ferguson, Jesse Patrick. Rocking Cosmopolitanism: Don McKay, Strike/Slip and the Implications of Geology. *ESC*, Vol. 32, Nos. 2–3, June–September 2009, pp. 165–187.

Forster, Sophia. Don McKay's Comic Anthropocentrism: Ecocriticism Meets "Mr. Nature Poet." *Essays on Canadian Writing*, Vol. 77, 2002, pp. 107–35.

Forster, Sophia. "Inventory is useless now but just to say": The Politics of Ambivalence in Dionne Brand's *Land to Light On. Studies in Canadian Literature*, Vol. 27, No. 2, 2002, pp. 160–182, https://www.proquest.com/docview/214494135?accountid=14593&parentSessionId=pcE2vciqeovqFapeon8j1GUariGRdK6D0hV2AWCMBng%3D&sourcetype=Scholarly%20Journals.

Hutchcroft, Blythe. Here at the Brink: Don McKay's Edge Poetics and the Articulation of Wilderness in Canadian Poetry, https://escholarship.mcgill.ca/concern/theses/t148fk447.

Lousley, Cheryl. Dionne Brand's Environmental Poetics. *Topia*, Vol. 34, pp. 39–61.

Lousley, Cheryl. Witness to the Body Count: Planetary Ethics in Dionne Brand's Inventory. *Canadian Poetry*, Vol. 63, 2008, pp. 37–58.

McKay, Don. *Angular Unconformity: Collected Poems 1970–2014.* Goose Lane, 2014.

McKay, Don. Baler Twine. In *Vis à Vis: Field Notes on Poetry and Wilderness.* Gaspereau, 2001, pp. 11–33.

McKay, Don. Between Rock and Stone: A Geopoetic Alphabet. In *Deactivated West 100.* Gaspereau, 2005. pp. 35–73.

McKay, Don. The Bushtit's Nest. In *Vis à Vis: Field Notes on Poetry and Wilderness.* Gaspereau, 2001 pp. 79–106.

McKay, Don. Ediacaran and Anthropocene: Poetry as a Reader of Deep Time. In *Making the Geologic Now: Responses to Material Conditions of Contemporary Life*, edited by Elizabeth Ellsworth and Jamie Kruse. Punctum Books, 2013, pp. 46–54.

McKay, Don. From Here to Infinity (or So). In *The Shell of the Tortoise: Four Essays & an Assemblage*. Gaspereau, 2011, pp. 115–36.

McKittrick, Katherine. *Demonic Grounds: Black Women and Cartographies of Struggle*. University of Minnesota Press, 2006.

Meeker, Joseph. *The Comedy of Survival: Literary Ecology and a Play Ethic*. University of Arizona Press, 1997.

Nevins, Joseph. Dying for a Cup of Coffee? Migrant Deaths in the US–Mexico Border Region in a Neoliberal Age. *Geopolitics*, Vol. 12, No. 2, 2007, pp. 228–47.

Nixon, Rob. *Slow Violence and the Environmentalism of the Poor*. Harvard University Press, 2011.

Okoth, Christine. The Extractive Form in Contemporary Black Writing: Dionne Brand and Yaa Gyasi. *Textual Practice*, Vol. 35, No. 3, 2021, pp. 379–394.

Quéma, Anne. Dionne Brand's *Ossuaries*: Songs of Necropolitics. *Canadian Literature*, Vol. 222, Autumn 2014, pp. 52–68. https://www.proquest.com/docview/1799550446?sourcetype=Scholarly%20Journals.

Sandilands, Catriona. Acts of Nature: Literature, Excess and Environmental Politics. In *Critical Collaborations: Indigeniety, Diaspora, and Ecology in Canadian Literary Studies*, edited by Smaro Kamboureli and Christi Verduyn. Wilfred Laurier University Press, 2014, pp. 127–42.

Yusoff, Kathryn. *A Billion Black Anthropocenes or None*. University of Minnesota Press, 2018.

3 Imagining Extinction in Deep Time

At a critical turning point in Jeff VanderMeer's *Hummingbird Salamander*, the protagonist (who goes by the alias "Jane Smith") views a video of a now extinct black hummingbird. A pragmatic security company employee and suburban mom, "Jane" has become entangled in an effort to locate Silvina Vilcapampa, a missing heiress with ties to both ecoterrorism and the illegal trade of taxidermied endangered species, including a stuffed specimen of the noir hummingbird she leaves as a cryptic clue for Jane. Watching the bird's ghostly trace on a burner phone, Jane is gripped by powerful feelings of wonder, admiration and loss:

> This creature that was everything I was not. We could not be more dissimilar, and yet, inside, I felt a welling up of sympathy for the toughness. For the miracle of this creature. Was this a gift Silvina gave me, too? How the world opened up? How it kept opening up.
>
> [...]
>
> Hadn't the hummingbird been a kind of miracle? Hadn't it diminished us not to see this as a miracle and protect it?
>
> (45–6)

For Jane, encounters like this become a catalyst that "opens" her up to perceiving what is at stake in mounting extinction rates of non-human species due to habitat loss and climate change, both in the wild and closer to home. Jane comes to see her own suburban neighborhood as habitat for more-than-human neighbors, such as beavers who have adapted to shrinking spaces, even as other species have been permanently displaced. Hence, Jane's admiration for the "toughness" of species capable of surviving in increasingly challenging Anthropocene environments is in tension with her growing realization that our world is irreversibly "diminished" with each additional loss.

We are accustomed to thinking about deep time in geological terms, a way of understanding temporality that properly applies to earth processes such as tectonic plates, volcanic eruptions, igneous rock, mountain building and erosion that transpires over billions of years—the lithic agency that is the primary focus in Chapters 1 and 2. From such a perspective, the average lifespan of an individual organism (whether a hummingbird or human being) occurs in the blink of an eye and is

DOI: 10.4324/9781003243946-4

gone; and even the average duration of a species (approximately 2 million years) may, in the context of such unimaginably long temporal scales, seem relatively brief, underscoring the vulnerability and ephemerality of living things. Yet as Rob Dunn reveals in *A Natural History of the Future*, lifeforms are being identified that seem to problematize neat distinctions between inorganic and organic temporality. Microbes have been discovered which thrive in the Earth's crust below the oceans, deriving energy from chemical reactions and undergoing extremely slow cell division such that "a single generation might take from a thousand to ten million years" (29). Equally remarkable is the fact that such microbes are estimated to comprise up to 20 percent of all the Earth's biomass, making their bizarre lifestyle "actually more common, whether measured in terms of biomass or of diversity, than is the mammalian lifestyle or the vertebrate lifestyle." The ubiquity of such astonishingly long-lived lifeforms, along with extremophile adaptations exhibited by many species of bacteria and archaea that, as branches of the "big evolutionary tree of life" are far more numerous than eukaryotic plants and animals, has led evolutionary biologists such as Sean Nee to conclude that, despite dramatic increases in human-caused extinction rates among multicellular species, *life on Earth* is, strictly speaking, by no means at risk in the Anthropocene era. Paraphrasing a 2023 presentation Nee gave at a conference on extinction in Okazaki, Japan, Dunn discusses the scientist's controversial position that

> the worst things we can imagine doing to Earth—nuclear war, climate change, massive pollution, habitat loss and all the rest—may affect multicellular species like us but are unlikely to lead to the extinction of most major lineages on the evolutionary tree.
>
> (238–9)

Within this broader context, the question arises: how concerned should we be that our species is currently perpetrating what Elizabeth Kolbert (and others) have identified as the sixth major extinction event on the planet?[1]

One could derive a sense of comfort, as Dunn seems to at times, that the "miracle" of life lies in its resilience or ability—despite the worst anthropogenic impacts—"to keep opening up" in unpredictable ways as Jane perceives in *Hummingbird Salamander*. But the risk of such a response to what might be called *organic* deep time is an attitude of Olympian detachment (and political defeatism) similar to that which I discussed in Chapter 1 with regard to geological deep time. If we conclude, as Dunn does, that "Nature is not under threat … and not ending anytime soon," we might be tempted to overlook one of his book's most important insights:

> what is under threat, what Bill McKibben announced the end of, are the lifeforms that we most relate to and that are most integral to our own survival. What is threatened are the species we love and the species we need.
>
> (239)

In other words, what we risk losing are our closest *kin* on the tree of life, organisms like Jane's hummingbird who, despite being "dissimilar" from humans in their marvelously strange adaptations, also inspire "sympathy" as fellow time-travelers who (to this point) have survived as awe-inspiring solutions to the challenges of existence.

Indeed, Dunn's insight here is less biological than ethical, implying a very different way of understanding the implications of anthropogenic extinction in the context of deep time than Nee's starkly scientific analysis. From the standpoint of our relationship with our closest evolutionary kin, current mass extinctions represent an irreversible rending of complex webs of life that, once gone, cannot be replaced: a loss that will, like ripples from a stone tossed into water, only magnify over time. From this perspective, mass extinction, like climate change, stands to be one of our species' most enduring and deleterious impacts if we do not act to reverse, or at least slow, such trends. As David Farrier puts it in *Anthropocene Poetics*, "perhaps more than any other environmental crisis, extinction pitches us into deep time: into awareness of the richness of our inheritance from the deep past, and the depleted legacy we will leave to the deep future" (92). How might Anthropocene citizens develop an ethics or a politics capable of honoring what Farrier calls the our rich biological "inheritance," or what might more properly be called conditions of possibility for continued co-existence with our eukaryotic kin such that microbial extremophiles aren't the only inheritors of the Earth?

Ursula K. Heise's *Imagining Extinction* explores ways that contemporary narratives about extinction operate in a variety of mediums (literature, graphic novels, documentary films, databases), some of which explore emergent ideas of "multispecies justice." Heise defines the latter as discourse that

> puts questions of justice for both humans and nonhumans front and center, even as it emphasizes that justice itself has to be imagined at the intersection of different cultural perspectives that may diverge in their conception of what is just.
>
> (202–3)

In particular, Heise is interested in extinction discourse that "move[s] environmentalism beyond the stereotypical narrative of the decline of nature without turning it into progress boosterism" (12), or what might be characterized as the Scylla of elegiac representations of lost nature on the one hand and the Charybdis of ecomodernist cruel optimism[2] on the other. Following Heise's lead, I will examine how extinction and ideas of "multispecies justice" are addressed in three works of recent climate change fiction: James Bradley's *Ghost Species*, Jeff VanderMeer's *Hummingbird Salamander* and Karen Russell's short story "The Ghost Birds." I'm interested in how each of these texts explores technological and affective responses to the problems of extinction and climate change. Bradley's *Ghost Species* deconstructs optimism regarding de-extinction as an intervention strategy, illustrating how "radical re-wilding" is shaped by a substitutive logic of capitalist consumption along the lines critiqued by theorists such as Catriona

Mortimer-Sandilands and Ashley Dawson. In contrast, VanderMeer's *Hummingbird Salamander* illuminates ways that grief can catalyze the desire for multispecies justice and resistance to extinction and illegal taxidermy trafficking. However, despite Silvina's attempts to break with such commodified responses to species loss, the novel ultimately reveals the failure of her political project, ways her idiosyncratic vision cannot achieve traction in an increasing apocalyptic, late-capitalist society. Instead of participating in collective, large-scale activism to address problems such as extinction and climate change, Silvina becomes increasingly isolated once her "Unitopian" commune fails, eventually resorting to bioterrorism, and finally to technological experiments aimed at re-engineering human perception as a last-ditch effort to force environmental adaptation. If Bradley's novel dramatizes the failure of ecomodernist technology as a strategy for mitigating extinction and VanderMeer's explores the limits of politics premised upon elegiac mourning, Russell's charts a third approach to imagining multispecies justice within the context of extinction. Valorizing neither techno-managerial optimism nor eco-melancholic despair, Russell's text offers a vision of the future that brings these discourses into conversation in ways that resonate with Dunn's evolutionary perspective. I will argue that Russell deploys a future-oriented eco-melancholia as a means to explore the potential of co-adaptation as a basis for building "a better world," or at least one that is not doomed to repeat historical mistakes and end in ecological apocalypse.

Living Absences as Venture Capital: Bradley's *Ghost Species*

In "An Ocean and an Instant," Australian novelist James Bradley explores three interrelated losses: of his remembered birth-place, of his father and of species via extinction, drawing a remarkable conclusion:

> Grief teaches us that time is plastic. A lifetime is an ocean and an instant. It does not matter whether something happened a week ago, a year ago, a decade ago: all loss is now. Grief does not stop, or disappear. It suffuses, inhabits us. The dead are both gone and never gone, living absences we bear with us.
>
> Perhaps something similar is true of extinction. What is lost remains with us, felt in its unpresence. The space between megafauna extinctions of 13,000 years ago and now is just a heartbeat in geological terms, the collapsing waveform of disaster we inhabit the convergence of a past and a present we can no longer escape.

For Bradley, the current acceleration in anthropogenic extinction rates (whereby "humans have killed at least two thirds of the world's wild animals" over the past 40 years) is a legacy of European colonialism extending back to the sixteenth century, and, long before this, can be traced back to pre-historic humans' likely role in global megafauna extinctions 13,000 years ago. In this essay, he suggests that the current extinction crisis represents an opportunity to finally come to terms with what should be our "grief" at losses we as a species have perpetuated on the living world. Yet, he also equivocates about the ultimate drivers of anthropogenic extinction: whether it is best understood as a product of capitalist colonialism (i.e. an historical, culturally specific problem best addressed by establishing greater socio-environmental justice) or whether it stems from something hard-wired in the evolution of our species.

Although such tensions are also at work in his recent novel *Ghost Species*, Bradley is less ambiguous on one key point: namely, that de-extinction projects do not represent a solution to the problem of anthropogenic extinction, but are in fact thoroughly ideological in their attempt to commodify species loss and to disavow experiences of loss and grief that might otherwise enable genuine disruption of ongoing anthropocentric practices. The novel focuses on tech billionaire Davis Hucken's schemes to restore extinct megafauna (such as woolly mammoths and aurochs) to huge "parks" around the globe in order to combat climate change and a related plan to re-create Neanderthals. Ostensibly, the logic that guides Davis's[3] vision is that in order to reverse climate change we need to "reconstruct entire ecosystems" (18) that existed before megafauna extinctions, particularly large swaths of grassland capable of reflecting heat, rather than absorbing it as dense forests tend to—thereby reducing the "albedo" effect of global warming that accelerates melting permafrost and methane emissions. Restoring large herbivores whose grazing kept forests from taking root is central to this "reconstruction" plan, as is the selective re-introduction of certain large predators like wolves and early hominins. Although Davis's project may initially strike readers as the stuff of "Jurassic Park" fantasy, his vision is actually in keeping with some real-life "radical rewilding" projects and with ecomodernist technocratic philosophy more generally.

As Emma Marris outlines in *Rambunctious Garden*, proposals for "Pleistocene Rewilding" (such as "Re-wilding North America" by field ecologist Josh Donlan, Dave Foreman and other conservationists) optimistically imagine that genetically engineered proxies of extinct megafauna could be introduced into the wild as part of an effort to re-create environments akin to those that existed before people came to North America (62). It should be stressed at the outset that such de-extinction schemes differ significantly from re-wilding projects aimed at re-introducing *still existent* species into locales where, over time, they became extinct (typically due to human causes such as over-hunting or habitat loss). Many re-wilding projects have focused on restoring species that have had positive net effects on regional ecosystems (wolf re-introduction into Yellowstone being a famous example) and the re-wilding movement as a whole has contributed in important ways to preserving global biodiversity.[4] However, projects such as "Pleistocene Park" differ significantly from mainstream re-wilding in that, instead of targeted re-introductions designed to assist nature's own power to restore healthy ecosystems, they imagine intensive technical intervention in order to create "designer" ecosystems wholesale. Although such projects may, at first glance, appear to be backward-looking, nostalgic attempts to restore a lost Edenic past, they are in fact better understood as future-oriented attempts to fully engineer what biologist Jens Christian-Svenning calls "top-down trophic interactions and associated trophic cascades to promote self-regulating, biodiverse ecosystems" (G78); or, as Marris puts it, Pleistocene re-wilding advocates have "really begun the project of designing brand-new ecosystems" (58). Indeed, as Heise notes in *Imagining Extinction*, the radical re-wilding advocated by ecomodernists such as Stewart Brand is driven less by a sense that "de-extinction makes any significant contribution to restoring ecosystems of the past" than it is by a sense that it is "a fascinating biotechnological experiment that might play a role in creating future ecosystems" (211). Though ecomodernists like Brand might see this as a positive step insofar as it transmutes "tragic and elegiac" stories about species loss into something they believe is more hopeful (i.e. accepting the necessity of human-engineered ecosystems), Heise is right to ask whether this does, in fact, "change the story" (210) we have been telling ourselves about extinction and about nature more generally. I would argue that *Ghost Stories* is a sci-fi-inspired narrative that ironically underscores how little ecomodernist de-extinction narratives change all-too-familiar stories of entrenched anthropocentrism, technological hubris and the reduction of non-human life to endlessly replaceable and fetishized commodities. Bradley shows us the extent to which Davis's de-extinction project represents a profound disavowal of extinction's finality and, as such, inoculates him and other like-minded technocrats from potentially transformative grief, such that, in a reversal of Bradley's meditations in "An Ocean and an Instant," species loss is *never now* but is forever relegated to an "elsewhere."

Right from the outset, it is clear that Davis's philosophy in the novel is consistent with ecomodernism. As he explains to Kate Larkin, a genetic engineer he is attempting to hire and who will become one of the novel's primary protagonists, he believes that many environmental problems stem from the fact that "old binaries, nature and technology, the human and the environmental, no longer make sense" and that "we have to accept … that it's up to us to re-engineer the world" (14). Davis responds to climate change and its cascading effects (including large-scale extinction) with an essentially technological solution: re-engineering ecosystems to make them more resilient, creating a "new nature, one in which humans play an active part" (17). He clarifies that piecemeal techno-fixes are no longer sufficient; it's not enough to restore extinct species or individual animals, but instead "we [now] need to reconstruct entire ecosystems" (18). Davis avoids devolving into a Cartesian cliché insofar as he realizes that his program of "entirely re-engineering our relationship with nature" (19) also requires us "to reimagine ourselves, the way we think, the sorts of attitudes that have got us where we are," including "the idea that we're distinct, separate, unique." Enter the need for resurrected Neanderthals as an intelligent "other," a hominin from whom humanity could learn how to live more sustainably. As Davis puts it:

> the world needs them. Look at the Earth, at what our carelessness has done to it. We can't let that happen again. We need to be tested by other minds, other perspectives. We need to learn how other eyes see the world. Think what we could learn from them, from other minds! Imagine speaking to another species! (20)

Davis's rationale for de-extincting Neanderthals may be admirable from an ethological perspective (i.e. a sense that we can learn from other intelligent species), but, as with other arguments for his project, it also belies a profoundly ahistorical, western bias. Davis's demand for a "new nature, one in which humans play an active part" is shaped by the binary thinking he (purportedly) seeks to escape—a binary that underwrites a specifically western, capitalist and colonialist praxis. Such praxis conveniently elides the fact that indigenous peoples "played an active part" in ecological management centuries before Europeans colonized the New World and regions such as Africa, Australia and island nations, an erasure that allowed colonists to equate native peoples with "nature" as exploitable resources. Likewise, Davis ignores the fact that, historically, western subjects (his assumed "we") had abundant opportunities to "learn how other eyes [saw] the world" had they respected indigenous ways of knowing instead of subjecting such "others" to slavery or genocide.

If Davis possessed an historically informed perspective, he would realize that the Neanderthals he wishes to bring into the world to serve as a bridge between humanity and other species would more than likely be "othered" by members of the socially elite class he represents: specifically, that they would be racialized, animalized and silenced in ways that parallel the historical treatment of people of color. Indeed, Bradley shows readers how the Neanderthal Eve, who is engineered for Davis's project, is repeatedly discriminated against on the basis of her species, race, gender and ability. This first occurs early in her life when, as a child, she's tormented on a playground by a group of white kids (three aggressive boys and a girl bystander) who mock her using the racist slur "monkey girl." Later, this pattern repeats when, as a teenager, Eve is rejected romantically because of her "ugly" facial features and "clumsy" body (194), her "broad shoulders and heavy musculature" (178) deemed an unfeminine liability due to sexist stereotypes. And although she is quite intelligent, her cognitive difference from humans is stigmatized even after she attains adulthood, as when she overhears two sapiens women speculating that she must be must be "on the spectrum" because of her limited "mental capacity" (249). Instead of being valued for her more-than-human perspective on the world, she comes to see herself as "monstrous," identifying with *Frankenstein*'s creature and the androids in *Blade Runner* (179) as similarly "othered" products of artificial engineering who are neglected by their creators (Davis leaves Eve's parenting entirely to Kate, given his sexist assumption that she will happily fulfill the role of a surrogate mother, having lost a child of her own.)

Ultimately, Kate concludes that the Neanderthal project she has been working on is "wrong because it fails to see that [Davis's team's] solution is part of the problem, a misplaced belief that this is another problem they can manage, engineer, and control" (64). In other words, despite Davis's efforts to the contrary, his techno-managerial response to climate change and extinction ends up replicating "the sorts of attitudes that have gotten us where we are" today. How is this possible, given Davis's technological brilliance and scientifically informed understanding of ecosystems? The novel suggests many possible answers to this question, but three stand out as especially significant from the perspective of debates about the Anthropocene. The first answer has already been alluded to: namely, Davis's ahistorical and culturally blinkered understanding how/why nature and non-European peoples have been "othered" within western, colonialist discourse and (consequently) his attempt to "engineer" solutions to complex environmental challenges without considering social justice issues.

A second answer to why Davis's project ends up failing is the role that capitalism plays in shaping his outlook and his corporate ventures. Although neither Davis nor Kate explicitly address this, capitalism acts as an invisible forcefield that shapes all of Davis's holdings, from Gather (the social media network that made him a billionaire) to the Hucken Foundation (a charitable organization that sponsors a range of environmental programs, from seed banks and genetic repositories to coral restoration and Pleistocene re-wilding). Given the scandals that erupt surrounding Gather's unethical data harvesting and sharing of information with Russian and Chinese intelligence services, one suspects that the Hucken Foundation's "charitable" environmental projects may, on at least some level, serve as PR green-washing. Likewise, as a pre-requisite to working for Davis, Kate must agree that her research "belongs to the Foundation" (33) and must sign non-disclosure agreements to ensure that the project remains secret to the general public and from competing corporations. Given that Davis regards Eve as an experimental prototype (i.e. a research product), this strongly suggests that he regards her as something that "belong(s) to the Foundation" and will seek a patent on her as a newly created transgenic organism (Eve's genetic makeup is partly sapien to permit fetal development within a human surrogate). Such patenting of lifeforms for the sake of profit is already well-established in the case of GMOs. As Ashley Dawson observes in "Biocapitalism and De-Extinction":

> The fundamental problem with de-extinction is that it relies on the throughgoing manipulation and commodification of nature and, as such, dovetails perfectly with biocapitalism. U.S. lawyers have already begun arguing that revived species such as a mammoth would be "products of human ingenuity" and should be eligible for patenting … De-extinction provides a mouthwatering opportunity for a new round of capital accumulation based on generating, and acquiring intellectual property rights over, living organisms.
>
> (179)

The same patenting could apply to other de-extincted transgenic organisms Davis produces, from mammoths to thylacines to giant sturgeon.

Moreover, Davis's methods of acquiring habitat for such animals is also fully in keeping with global neoliberalism. For example, during a Steve Jobs-style press conference in Siberia highlighting his Foundation's latest work, Davis announces that he has "negotiated an agreement with the Russian government for the control of almost two million hectares of state land in areas characterized by continuous and semi-continuous permafrost and forest" (53), ostensibly to study the ecological effects of global warming and de-extincted animals. The privatization of such huge areas of previously state-owned land raises a number of ethical questions, including whether working-class laborers, indigenous peoples and/or wildlife have been displaced to make room for Davis's large-scale "park." The fact that aurochs will be the first animals restored to this vast area gives the project a totalitarian tinge, insofar as the Nazis spearheaded a breeding program to restore this species. Finally, it's not impossible that Davis's vast Pleistocene parks in Russian and Lithuania could, one day, have been converted into tourist attractions had world governments not dissolved due to out-of-control climate change following the collapse of the West Antarctica ice sheet. Although the proposed future "Pleistocene Park"[5] in Eastern Siberia and its "radical geoengineering scheme" to halt melting permafrost by re-introducing megafauna (including de-extincted mammoths) is the clearest real-world model for Davis's project, other examples of vast private parks dedicated to re-wilding already exist and some are tourist attractions. Media mogul Ted Turner's 360,000-acre Armendaris Ranch in New Mexico is a good example of the latter. Significantly, this ranch hosted a conference for Pleistocene re-wilding enthusiasts including Paul Martin, Josh Donlan, Michael Soulé and Dave Forman in 2004, which culminated in the publication of the "Re-wilding in North America" paper in *Nature* the following year. Today, affluent visitors to the ranch can experience re-wilded endangered species, such as the 110-pound Bolson tortoise or African oryx, while staying in a luxury hotel with a spa and geothermal hot springs, or can rent another Turner property in Northern New Mexico called Vermejo Park reserve (also home to reintroduced endangered species such as cutthroat trout) for $1,800 to $9,000 per night. All of which is to say that, although Davis doesn't foreground commercial reasons for his Foundation's activities, his excessive wealth no doubt shapes both his vaunting ambition and his over-confidence in what can be achieved via technological means. When Kate asks her partner (and fellow scientist) Jay whether Davis really believes that "the world can be re-engineered," Jay's response sums up his employer's financially conditioned hubris: "When you have his kind of money almost anything is possible" (24).

Finally, Davis's privilege shields him not only from having to contemplate ethical problems that arise from his business and philanthropic ventures but also from having to engage with existential realities such as death and loss that are a major dimension of extinction. This is a third major reason that his radical re-wilding projects fail. On an intellectual level, Davis appears to be extremely concerned about climate change's cascading destruction, from catastrophic fires that scorch Arctic tundra to melting sea ice to greatly accelerating extinction rates. On an emotional level, however, he seems oddly detached from these losses, never expressing any sense of grief about the living world collapsing around him. One could see this as being in keeping with a general flatness in Davis's affect, which, combined with his frequent incomprehension of social queues and tendency to "rehearse" conversations in advance perhaps indicates that he is neuroatypical. After all, he seems equally unmoved when Eve is born, displaying no happiness that his historic endeavor to bring a Neanderthal-sapient into the world succeeded, nor any affection for his creation (Kate notices that "at no point did Davis hold or touch the child" (60)). Yet a less ableist explanation for the tech guru's apparent lack of sadness about mass extinction is that capitalist biotechnology has provided him a means to circumvent extinction's terrible finality, converting irreparable loss into an opportunity to create new commodities that "replace" their lost genetic antecedents. Catriona Mortimer-Sandilands's analysis of eco-melancholia versus commodified mourning in "Melancholy Natures, Queer Ecologies" is a helpful framework for illuminating the substitutive logic that shields Davis from experiencing a sense of loss. Inspired by theorists such as Sigmund Freud, Bruce Braun and Judith Butler, Mortimer-Sandilands argues that "experiences of environmental loss … are present, tangible" and increasingly common in a world ravaged by climate change and capitalist development, yet, because modern western societies lack the political will, language or rituals to mark such losses, they remain officially "ungrievable." Nonetheless, mourning for such losses "can be easily commodified … and indeed … the marketing of environmental loss can be a positive boon for the unfettered progress of capital" (338). Although Mortimer-Sandilands cites ecotourism and fetishized "disappearing" wilderness as primary examples of this, de-extinction can also be seen as a potent example of how painful loss can be converted into compensatory commodities that are "not only infinitely consumable, but [also] infinitely replaceable." Instead of allowing himself to be haunted by grief for pre-historic as well as current species losses, Davis dispels these ghosts via the apotropaic magic of biotechnology. But in doing so, he renders extinction both impermanent and meaningless.

What could have been learned from past and present anthropogenic extinction events has been rendered into a televised spectacle of technological marvels. At one point in the narrative, Kate is moved by a report on one of Davis's parks in Lithuania, thinking: "The mammoth are beautiful: vast russet mountains that tower over the landscape and sway slowly as they walk, the wonder of them undiminished by the decade that has passed since they were first created" (168). Yet she goes on to note that "what the report didn't reveal is that Davis's efforts seem to be making things worse rather than better" (169). Kate knows the "reforestation programs have not been working" and that mortality rates in many animal populations have risen around the globe, due to habitat loss, pesticides, warming waters and other (unknown) factors. As Mortimer-Sandilands clarifies, a more politically meaningful response to such losses would be an eco-melancholic one that insists that non-human extinctions *should be grievable*, that the lives of more-than-human subjects (like the lives of queer folk) matter and deserve to be mourned when they pass away. Such melancholia "represents a holding-on to loss in defiance of bourgeois (and capitalist) imperatives to forget, to move on, transfer attention to a new relationship/commodity" and thereby "holds [open] the possibility for memory's transformation into ethical and political reflection" (354).

Suggestively, it is Kate and Eve that open themselves most to eco-melancholic grief in the text rather than Davis. When global ecological catastrophe spirals out of control in the novel's final chapters, Davis retreats to "a safehouse in New Zealand," occupying a defensive redoubt that literalizes his psychic and emotional isolation throughout the narrative. In contrast, Kate has long been "haunted" by species extinction and a sense of "accelerating collapse," though she acknowledges that there's also something "numbing about this process" (169), something that's difficult to talk about with her scientific colleagues. Ultimately, it is the neo-Neanderthal Eve who most vividly experiences eco-melancholic grief in the text but also potentially transformative hope that she (and others like her) will give rise to a different future. Stumbling upon a mammoth corpse while journeying in France on a quest to find other transgenic Neanderthals that she had seen online, Eve is overcome not just by the visceral horror of the decaying beast but also by the dead forest that surrounds it. A place that was once animated by deep webs of information shared over eons (a "wordless remembering") is "now dead, the smell of it lingering like the stink of the mammoth carcass." Overcome by grief for this symptom of larger ecological devastation, Eve asks:

> Is there a word for that loss? A word that might name this rupture in the world? A word that might capture the way that all that has happened has sundered this place from its past and left it storyless and alone?
>
> (264)

The novel's concluding pages suggest that by accepting and holding onto ecological loss, Eve can also accept life and begin the process of choosing another path forward that (like the planet) "slip[s] free" of older sapient "fantasies of control" (265). As she passes through ruined cities and towns, she notices that "in the lands outside them life is changing, growing wilder," and that (like her) other transgenic beings like the mammoths have migrated far from their intended reserves. Though many surviving human beings live now live in caves, "history run [ning] backwards" such that their "beginning becomes [their] end," Eve does *not* return to the Neander Valley where the bones of her ancestors were first identified as being distinct from *Homo sapiens.* Symbolically, this suggests that rather than merely repeating history, neo-Neanderthal experiences of ecological grief may afford her people a different future trajectory. Telling stories to her own child in the opening chapters, Eve "draws filaments from the darkness and weaves them together to create meanings, purpose, shape, arranging the elements to reveal the world or perhaps to make a new one" (1).

Hummingbird Salamander and the Limits of Eco-melancholia

For all its potential as an affect theory that tries to create a language for ecological grief and a means to mourn the loss of more-than-human entities, eco-melancholia is best understood as a possible catalyst—not a guarantor—for progressive political action. Although it has enormous value in disrupting the substitutive logic of commodified mourning in late capitalist societies, environmental grief is also analogous to an unstable chemical element: it can be transformative, but is also highly unpredictable in its effects. "Dark ecological" emotions such as eco-melancholia can inspire us to care for devastated landscapes and to mourn the ghosts of past and present extinctions, but they can also lead to intense despair, anger and/or cynicism, particularly in light of inadequate political responses to existential threats such as climate change in countries such as the United States. In *Staying with the Trouble*, Donna Haraway discusses two problematic responses to "the horrors of the Anthropocene and the Capitalocene," namely, naïve hope in technofixes on the one hand and "game over," apocalyptic despair on the other. Suggestively, it is the latter that Haraway claims is "harder to dismiss" than the former because the "sublime despair" at work in some varieties of abstract futurism can readily devolve into a "politics of sublime indifference" (4). Although my analysis of *Ghost Species* focused on the emotional avoidance underwriting Davis's hope that de-extinction could constitute a "technofix" to problems such as species loss and climate change, profound despair can also lead to psychic isolation of a different, equally destructive order. In this section, I will explore the risks of grounding politics in environmental grief by analyzing Silvina Vilcapampa, the remarkable eco-melancholic protagonist in Jeff VanderMeer's *Hummingbird Salamander* (2021). Unlike Davis Hucken in *Ghost Stories*, Silvina's profound grief and anger at mass extinction is a primary catalyst for her environmental activism. Indeed, her eco-melancholic refusal to numb herself to ecological destruction and species loss shapes her trajectory from suspected eco-terrorist to utopian cult leader to would-be bioterrorist. Yet because (as the daughter of a billionaire) Silvina shares Davis's deeper internalization of neoliberal individualism and the values of self-reliance, she remains an isolated figure, waging a largely one-woman war against the system. Thus, despite her emotional differences with Davis, her ecological projects ultimately fail in very similar ways, suggesting the limits to affect as a basis for environmental praxis. Like Davis, Silvina ultimately fails to embrace collaborative world-building as a basis for comic survival. Withdrawing with disgust from flawed humanity once their utopian schemes falter yet also unable to experience intimacy with others, both protagonists fail to grasp that, in Haraway's parlance, "[s]taying with the trouble requires making oddkin" and effective praxis entails "becom[ing] with each other or not at all" (4). Perhaps as a consequence of her refusal to accept an imperfect politics grounded in such earthly collaborations, Silvina has what can only be called a *spectral* presence in *Hummingbird Salamander*. That is, while she is experienced as a powerful ghost by the narrator "Jane" she also remains otherworldly—at a remove from quotidian habits or ordinary human relationships. "Jane" can't "imagine Silvina having lunch or playing a board game" (160), but also can't picture her in loving relationships with human or animal companions, despite her passion for ecological causes. Silvina sums up her spectral identity in a journal entry that foreshadows her eventual recourse to bioterrorism: "We're ghosts trapped in the wreckage of our systems. So why shouldn't we haunt them? Why should we not avenge ourselves upon them? Why be merciful?" (157).

As I will illustrate, *Hummingbird Salamander* dramatizes both the power and the limits of eco-melancholic despair, showing environmental grief's capacity to spark Silvina's utopian dreams but also reinforcing her sense of alienation from other human beings such that she eventually pursues a fatal messianic quest to single-handedly save the world via biotechnology.

Significantly, one of the first things that readers learn about Silvina is that she "want[s] to be lost." During a journey to Quito, Ecuador that she undertook at the age of 21, she states

> I want to be so far beyond anything that there is no map, and the compass spins wild. And when I come back, if I come back, you need to know I've changed, and with that change it means I carry "lost" with me everywhere, even in the heart of the city. That I am lost forever and that's how we need to be. So systems can't find us, can't wreck us. So our heads are clear.
>
> (24)

Following nineteenth-century naturalist Alexander von Humboldt's route, Silvina pursues a classic Romantic quest to lose herself in wilderness, fleeing from her acute sensitivity to crowded urban places but also motivated by a desire to "clear her head" in order to "know the world better … how it truly is" (25). Ironically, though she seeks to escape from oppressive modernity and meets with some indigenous activists in Ecuador, what she primarily encounters in remaining rainforest is evidence of "Oil companies and mining [ventures that] had devoured part of the route—leading to a revelation that "she couldn't run. Couldn't hide. What she hated would always be in front of her." It is no coincidence that Silvina's father Matias Vilcapampa amassed a fortune from the very extractive businesses that she abhors: namely "mining, oil and other dirty industries" (66) and that he engages in PR greenwashing by outsourcing such industries to nations outside of Argentina, his country of origin. Domestically, "as if not to soil the nest" Matias sticks to his family's traditional "cattle ranching" business, and "even gave to environmental organizations." Due to such philanthropy, he is regarded as a "pillar of [his] community" (68) and as a visionary who "believes in the future" even as his enterprises actually contribute to climate change that "[takes] the future away." As a member of her family's inner circle, Silvina is acutely aware of its hypocrisy and is appalled by her father's pride in his capacity to dominate and exploit the natural world, symbolized not only by his profiting from extractive industries but also by his big game trophy hunting and, readers later learn, his role in illegal wildlife and taxidermy trafficking.

Indeed, horror and grief at her family's role in wildlife extinction is Silvina's gateway to becoming an activist. Following her travels in Ecuador, Silvina becomes an animal rights activist who is repulsed by "how many dead things haunt us in our daily lives … as purses, handbags, shoes—even as heads on walls" (100). Her inability to repress an awareness of ubiquitous animal sacrifice (to render it as the "setting" of everyday life) eventually also affects "Jane" as she unravels the mystery of Silvina's secret activities and purported demise. After Ecuador, Silvina is arrested for an alleged conspiracy to murder five wildlife traffickers, but is released on the basis of inadequate evidence (53). Similarly, she later writes a manifesto for the "Liberation of the Earth at any cost. Liberty or death—death to those who oppose" and stands trial for her alleged involvement in an ecoterrorist bombing of Argentinian citizens but is, again, acquitted (48).

The novel leaves open the question of whether or not a youthful Silvina actually committed these crimes (despite her revolutionary rhetoric), and suggests that her prosecutions may have been exercises in public humiliation orchestrated by her father and like-minded patriarchs as retribution for her activism (she "reported her father's 'collector' friends to the police. At age twelve" (117)). Indeed, such an interpretation would be consistent with remarks that VanderMeer has made in interviews about *Hummingbird Salamander* regarding what he sees as a troubling rise in the criminalization of activism in the US in recent years.[6]

What the novel does make clear is that it is not until Silvina journeys to the Pacific Northwest wilderness in the United States and encounters endangered species first hand that the full weight of mass extinction and grief at the wider tragedy unfolding all around us truly radicalizes her. Immersing herself in the wild for close to two weeks (until animals that she encounters no longer appear to see her), she happens upon a naiad hummingbird drinking rainwater from a puddle on the ground. Her journal records her awe at seeing this tiny, jewel-like black bird engaged in a rarely witnessed behavior, and her acute anguish when she learns that the species is highly endangered:

> you can imagine how I felt when, later, I identified the bird and realized how rare the species was and headed toward extinction, and what I had borne witness to wasn't just a minor miracle but, in fact, a moment that would replicate only another hundred or another thousand times. That in the history of the world, a naiad hummingbird would only come to the ground to drink a finite number of times before they no longer existed. This thought was unconscionable to me. Unbearable. It ripped me in two. It destroyed me. And then remade me, and I became someone different than before.
>
> (127–8)

The transformative nature of Silvina's "unbearable" mourning here recalls Thom van Dooren's discussion of philosopher Thomas Attig's "understanding of grief as a process of 'relearning the world'" in *Flight Ways*:

> As we grieve, we appropriate new understandings of the world and ourselves within it. We also become different in light of the loss as we assume a new orientation to the world… Some of what we took for granted in ourselves or in our life patterns is no longer sustainable. Relearning the world thus requires that we make changes.
>
> (Attig quoted in van Dooren, 139)

Seeking to challenge grief as a marker of human exceptionalism, van Dooren discusses not only mourning behaviors in other species, but also ways that Attig's concept of mourning as a transformational process of "relearning the world" reflects complex webs of entanglement that transcend species divides:

> What grief points to here is a particular kind of *shared world* or *shared life* … a particular sociality rooted in our being *emotionally* at stake in one another's lives. This possibility, this way of being with others, is a complex biosocial achievement, requiring the coming together of evolutionary histories and emotional and cognitive competencies to produce embodied subjects who are unavoidably emotionally entangled with one another. It is only inside of these particular biosocial configurations that the passing of another out of the world can be experienced and felt as genuine loss.

Why is Silvina's encounter with the naiad hummingbird such a turning point in her life, and why does contemplating its immanent extinction radically "destroy" but also "remake" her identity? After all, her journal records that, as far back as the Quito expedition, "she had encountered the naiad hummingbird. Or had sought it out. Or had found them for sale in a local market as love charms, dead and dried out" (119).

While there are multiple reasons why the hummingbird's endangerment could have been experienced as impactful at this juncture in Silvina's life, the language she uses to describe her epiphany suggests that, perhaps for the first time, she "really *get[s]* … the multiple connections and dependencies between ourselves and these disappearing [non human] others" and/or "appreciate[s] all the ways in which we are at stake in one another, all the ways in which we share a world" (van Dooren 140). Here, Silvina seems to grasp that for all their differences, there are a number of striking similarities between herself and the hummingbird; both are beings which migrate great distances, not being rooted in a single place our nation; both are thirsty and require water to survive (she sees the bird drinking at a moment when she herself had "drunk very little water" all day); both are capable of astonishing things, but both are also vulnerable (the bird is "defenseless" as it sits on the ground; Silvina is a fugitive) to harm and to death (the bird is black, a color associated with mortality and that mirrors Silvina's "eyes so dark, they registered as black" (31)). Both belong to species whose "shared world" is jeopardized by global warming; by the time "Jane" receives a taxidermied specimen of the hummingbird as one of Silvina's final bequests, the bird "had gone extinct because of poaching, habitat loss and climate change" (41). The fact that the Vilcapampa family's wealth has been generated via practices that have directly contributed to these problems must have been intolerable to Silvina at this moment, given the outsized role that such praxis played in rendering evolutionarily unique beings like the hummingbird extinct. What's at stake here is nothing less than a rupture in the long "history of the world," the loss of "a part of what invents and maintains [the world] as world" (219) in Vinciane Despret's parlance. Just as the "world bursts from absence" with the death of a species, Silvina's identity is annihilated in grief, and re-born in rage. From this point on, whether or not she participated in ecoterrorism in the past, she will now devote the rest of her life to finding a way out of the "wreckage of our systems," even if that means becoming an angry ghost: *"Why should we not avenge ourselves upon them? Why be merciful?"*

Silvina's first major attempt to create a counter-cultural alternative to neoliberal hegemony is her Unitopia colony, a cult-like community made up of herself and a handful of devotees. Searching the abandoned remains of the Unitopian compound, Jane is struck by its "sweet, naïve quality" but also by the way it constitutes a monument to Silvina's futile utopian dreams of establishing an "independent" (182) way of life untainted by the outside world. Jane asks herself whether people had ever really lived in Unitopia and sardonically concludes: "How sad if they had worked here for a different future … but lived in a subdivision named 'Lake Woods' or 'River Creek.' Revolutionaries trapped in a kind of gently theme park life" (180).

Having re-learned that it isn't possible to "lose" oneself in an Anthropocene world where there is no "elsewhere," Silvina embarks on a second, darker "utopian" project, one that uses genetic engineering in order to produce a "chemical biomimicry" whereby human embodied perception would come to resemble that of other species. Jane describes this project as one that sought "a new, true seeing" whereby a person could "let the world in through [their] pores like a salamander, see all the colors of the flowers only a hummingbird could see" (345). Apparently, over time Silvina concludes that it's not sufficient to change people's minds via political manifestos or utopian tracts. Rather, what's needed are more radical changes to our embodied being such that instead of learning to value the world we "share" with animal others via grief (as in van Dooren's ethics), we *literally* (physiologically) see it anew through more-than-human perception.

Such a "solution" to the problem of human estrangement from the rest of the natural world is in keeping with VanderMeer's long-standing fascination with weird forms of embodiment that deconstruct human/non-human dualisms, from the transformative fungal contaminations in the Ambergris trilogy, to the "brightness" that infects the biologist and other characters in the *Southern Reach* trilogy. Alison Sperling's "Second Skins: A Body-Ecology of Jeff VanderMeer's *The Southern Reach Trilogy*" eloquently explores how the SR trilogy "maps Anthropogenic change … on and through the body" (250) such that "nonhuman elements are experienced anew once the [human] host has been infected, creating a complex new relationship with the world that complicates formulations of sickness, almost mak[ing] it appealing" (243). Similarly, Silvina infects herself with an unspecified bioengineered agent (possibly a virus), to serve as a test-subject for "ecstatic" contamination (345), but ends up an isolated corpse entombed in a compound that also houses an artificial "ark" (ecosystem) designed to preserve biodiversity in a world ravaged by climate change and species loss. One key difference in Silvina's bodily contamination and that of other figures in VanderMeer's oeuvre is that most of his human protagonists are acted upon by more-than-human agents that they do not control and/or find inexplicable or alien. Hence, Jane is somewhat ambiguous in her assessment of Silvina's attempts to use technology as a means *force* our species to change ("*we must change to see the world change*"), unable to definitively say whether this project reflects "derangement or genius" (345). Although she ultimately decides, in a leap of faith, to inject herself with what remains of Silvina's experimental serum because she believes the potential benefits of transformation are "worth the price, to change the world," she admits that "odds are, someone will find [her] dead on that hillside" and that "the ark inside the mountain is probably doomed, and Silvina, in the end, was just someone desperate and alone and delusional" (350).

Indeed, one thing that's unequivocable in the novel is the terrible cost of Silvina's final utopian ambitions. In order to finance her high-tech "solutions" to the problem of extinction she (ironically) ends up participating in the very wildlife trafficking and black-market taxidermy industries that bankrolled her father's enterprises, which is no doubt how she acquired the taxidermied naiad hummingbird that sparks Jane's initial investigation. Given the hummingbird's crucial role as a trigger for Silvina's grief and radicalization, it's hard not to read her eventual profiting from taxidermy and trafficking of endangered wildlife as a profound betrayal of her most fundamental ethical commitments—something which, in the case of the hummingbird, directly contributed to its extinction. Thanks to her father's involvement with "wildlife trafficking cartels [that] manufactured need [telling] those inclined to buy that this or that animal was good luck or the next hip thing for the newly rising rich," Silvina is intimately aware of the macabre (and real-life) illegal trade of taxidermied endangered species that are primarily purchased by the world's economic elite. Taxidermy haunts the entire novel as a melancholy, subliminal signifier of Anthropocene mass extinction. In one particularly visceral passage, Jane is literally buried under of mountain of taxidermy that the Vilcapampa family has stockpiled as inventory in a warehouse. Yet Silvina's horror and grief at this trade's environmental destruction doesn't stop her from becoming what she detests—an exploiter like her father—once she embarks on her monomaniacal quest to "save the world" via science.

As with Davis Hucken in *Ghost Species*, there's a messianic quality to Silvina's utopian schemes that's symptomatic of both overreach and extreme psychic isolation. Silvina's social isolation goes back to her youth, where she went to school "the way an army brat does: knowing she would be somewhere else the next month or semester," never bonding with a specific place or specific people but instead concluding that: "the first thing you realize is that you are all alone and you can rely only on yourself" (118). Likewise, Jane is disturbed by the fact that there's "never mention of any pets" in Silvina's journals nor "a single mention of a lover, a boyfriend, a girlfriend" (160). Silvina's emphasis on "self-reliance" and isolated agency (no doubt shaped by her family's capitalist values) eventually affects how she formulates her plans for ecological salvation. Instead of working with a variety of activist organizations on causes that she believes in, she elects to become the guru of a short-lived ecotopian community. Indeed, Jane surmises that Silvina's motives for establishing Unitopia were less altruistic than egoistic, a quest to overcome her sense of being "stateless," of "belonging … to no place and to no one. Perhaps, at first, Unitopia had felt like a way to create her own country" (181). After Unitopia fails, Silvina "goes dark" (166) and embarks on a perilous solo venture to engineer the transformative drug that eventually ends her own life. But there's evidence, in a "declassified top secret" US government report that

Jane reads, that if this drug had achieved its intended purposes it may have been employed as a biological weapon "with the goal of broad use against civilians" (168). If this report is accurate, it suggests that Silvina would not have hesitated to unleash her "magic elixir" (350) on an unsuspecting public, seeking to coercively achieve her ends without broader debate or consent. As with Davis, Silvina's plot underscores her social alienation and the risks of concentrating too much power in the hands of elite investors capable of deploying potentially hazardous technology without any public oversight. While Silvina is initially motivated by eco-melancholic grief in ways Davis is not, it is revealing that their quests end up in very similar places—i.e. as would-be messianic saviors bunkered in technological "safe houses" as the world around them careens into climate change apocalypse. Perhaps, in the case of Silvina, this is because (unlike VanderMeer himself) she didn't "have the patience to sustain" (182) collaborative enterprises such as community activism and re-wilding, ways to "stay with the trouble" that are "always situated, someplace and not noplace, entangled, and worldly" (Haraway 4). Although Silvina's goal of saving wildlife is laudable and her eco-melancholic capacity to "love" poisoned or damaged places is a sign that she's "still in the fight" (101), she never fully embraces a sympoietic *praxis* in order to achieve her ends, autopoietically isolated to the very end instead of *being* the change she seeks. Her journey from being an environmentalist to a bioterrorist illustrates that although she is "brave" enough to "let herself be haunted" (VanderMeer "Hauntings") by the trauma of mass extinction, there are also limits to eco-melancholia as a basis for effective ecological politics.

Future-oriented Eco-melancholia and Adaptation in "The Ghost Birds"

Cataloguing a growing list of species losses in her near-future world, "Jane Smith" confesses that fully acknowledging such losses would mean "putting a bullet in [her] head," so she learned to "care less"—a defensive withdrawal that Silvina Vilcapampa terms a "fatal adaption" (37) insofar as it represents an attempt to escape from Anthropocene realities via self-protective cynicism, but does nothing to intervene in cascading destruction. Karen Russell's recent short stories that address the interrelated problems of climate change and extinction likewise explore what it might mean to "adapt" to a world wherein humans and our evolutionary kin are gravely imperiled. Despite Mortimer-Sandilands's efforts to demonstrate that eco-melancholia represents an alternative to nature nostalgia and the capitalist commodification of ecological loss, there is a temporal dimension to melancholia that biases it towards the past—towards losses that have *already* happened, damage that has already been done, processes of destruction already underway. Thus, while there is a risk in fetishizing lost "pristine" nature, there is also a risk that we may (collectively) become *too adapted* to damaged landscapes, learning to love what's left in the wake of global ecological ruin but powerless to effect meaningful change to prevent future apocalypse. A brilliant student of mine once posed the question: is a *future-oriented* eco-melancholia possible? Is it possible to grieve for losses that have not yet happened but well may if we don't change processes already underway (such as climate change and extinction)? How might such anticipatory mourning inspire *transformative* adaptation, rather than bolstering neoliberal "sustainability" and other status quo models of the future? Might such mourning constitute an important alternative to the "fatal adaptation" that Silvina critiques, and to the acceptance of ongoing loss as inevitable?

In her short story "The Ghost Birds," Karen Russell creates a disturbingly familiar storyworld that imagines where our climate-changed world may be headed, particularly in regards to species loss. I believe this text is a striking example of future-oriented eco-melancholia, or anticipatory mourning, that presents readers with two very different adaptive responses to ecological endangerment: one premised upon the anthropocentric notion that humans can use technology to adapt to anthropogenic destruction and which is unconcerned about dooming other lifeforms to extinction, and the other which mourns anticipated mass extinction and seeks to resurrect lost ideals of social justice and combine them with a new politics of multispecies justice. Importantly, the former assumes that natural entities will either have to adapt to human needs or will cease to exist, whereas the latter understands adaptation as a mutual, open-ended and flexible process that emphasizes multispecies *co-adaptation* as a key to survival. In "The Ghost Birds"'s projected future (the US Pacific Northwest circa 2081), the former ethos has triumphed insofar as fossil-fuel capitalism has continued unchecked and the powerful, in their "infernal ingenuity," have developed technology that permits humans to adapt to catastrophic fires and other unnatural disasters, "find[ing] ways to survive the death sentence we had delivered to our gasping cohabitants of this planet" (58). While ecomodernist dreams of using large-scale geoengineering to mitigate climate change seem to have failed in this society, it nevertheless clings to a belief that technology is the best (and only) solution to ecological destruction, while the rift between economic elites and those exposed to ever-greater environmental risks has only widened and non-human extinction has accelerated. Yet this world is also haunted by traces of what could have been—evidence of a different path not taken that might have birthed a future wherein greater commitment to social and environmental justice may have mitigated climate change's deleterious effects and minimized multispecies suffering. Within Russell's near-future society, this haunting constitutes a tenuous hope that future life on planet Earth may rebound despite cascading apocalypse; but in our current historical moment, the text's anticipatory mourning challenges our generation not to let this worst-case scenario unfold, to act now to prevent future catastrophe. The eco-melancholic "ghosts" within the text are (thus) multifarious and strange, at once haunting reminders of past, irreversible extinction and harbingers of hope that such losses are not inevitable in the future.

Like Jeff VanderMeer's work, Karen Russell's published fiction to date has often engaged with our ecological crisis and combines realist and non-realist techniques to explore issues such as climate change and mass extinction. As Laura Shackelford and I explored in *Surreal Entanglements*, VanderMeer's "weird fiction" hybridizes elements of horror, surrealism, science fiction and non-fiction "nature writing" to address such concerns, and Russell likewise experiments with genre-defying magical realism in order to register trenchant psychological insights about her characters and how they subliminally experience the Anthropocene. And as with VanderMeer, Russell has clarified that she employs such stylistic innovations not merely as entertainment for her readers. In a 2019 interview for the *Chicago Review of Books*, Russell stated:

> Sometimes I fear people see the word "magic" and associate it with a kind of escapism. A friend of mine, the wonderful Porochista Khakpour, wrote that "great literature merges escape and confrontation." It's a paradox that sometimes you have to go to Middle Earth or Macondo or Dune to discover the truth of this world.

Significantly, Russell employs paranormal or supernatural elements in her world-building, but often in the service of commenting upon recognizably earthly and/or historical phenomena. In *Swamplandia*, Osceola Bigtree attempts to communicate with her dead mother via seances and falls in love with a ghost, but her home in the Everglades is also haunted by the historical traces of displaced Seminole peoples and decades of ecological mismanagement by settler colonialists. Stories in the *Orange World* collection likewise explore topics such as a woman's psychic possession by a Joshua Tree ("Bad Graft"), an aging farmer's nostalgia for buying and raising twisters ("Tornado Auction"), and a post-apocalyptic Florida populated by bat-humans and haunted by neoliberal capitalism ("The Gondoliers").

"The Ghost Birds" likewise combines elements of realism and the fantastic in order to explore interconnections between accelerating climate change and mass extinction. In a 2021 interview for *The New Yorker*, Russell explains that the story's ecologically devasted near-future setting in which smoke-filled air has become unbreathable without the assistance of respirators and all of the world's birds have perished was inspired by the author's first-hand experience of "our fire-lit reality in 2020–2021," including the deadly heat dome and massive wildfires that affected Portland in the summer of 2021. In this conversation, Russell expressed her "disorienting" sense that such ecological disasters are becoming normalized in the climate-change era: "maybe the scariest thing is that [such disasters] no longer felt to us like anomalies" but instead "seem like they might become seasonal events." Similarly, she cites reports about "mysterious die-offs of birds in multiple states" as another influence on "The Ghost Birds," though she also admits to being somewhat "behind the times" in her awareness of this problem. What might it mean to become "adapted" to such losses, and/or to become habituated to the ecological damage that such extinctions signify?

"The Ghost Birds" depicts just such an adaptation due to US society's failure to take action on climate change or to reform neoliberal capitalism in any meaningful way. In 2081, current social inequities have only become more pronounced and ecological damage is spiraling out of control: the rich inhabit domed compounds above the smoke line and have "colonized" the sky ("a single human being now claims to own all the sky that lifts from the Andes to Mars" (57)), while the most a lower-class person can hope for is to become "an indentured servant on the floating starships" that are part of "trillionaires' intergalactic fleets" (58), thereby escaping from the ruins of planet Earth. For the average person, affordable in-person education has become a thing of the past (only the wealthy can afford in-home tutors) and even "air has become damn expensive in the past decade" given the necessity of oxygen for life-supporting respirators. Because ecological instability has become commonplace ("phreatic eruptions," along with "food shortages, tsunamis, hurricanes, and wildfires" (57) happen all the time in this world) freedom of movement has been severely truncated by security forces ("Surveillers") hired by social elites who are "paranoid in proportion to the suffering that surrounds them." Bird-watching in this dystopian future—or rather, *ghost* bird-watching—becomes an act of rebellion, and is punishable by extreme force (Surveillers often execute birders caught "trespassing" on land they don't own).

Enter the absurd and strangely poignant obsessions of Jasper, the story's narrator and long-term member of the Paranormal Birding Society (PBS). A middle-aged divorcé with a teenage daughter named "Starling," Jasper and his fellow ghost-birders look for traces of deceased avians, hunting for signs of "real birds that have gone on swimming and singing beyond extinction." He and his fellow enthusiasts insist that

> if there is one myth the ghost birds have exposed, it's that death means stasis. The flocks we track continue to cross oceans and continents, and the Paranormal Birding Society has been collecting fresh data on their distribution patterns, undead coloration, and evolving calls and songs.
>
> (58)

So strong is the birders' drive to pursue such revenants that they're willing to put themselves in considerable danger: "Orrine was shot in the former Okefenokee Swamp, while searching for traces of the ivory-billed woodpecker … Popple lost his pinky to a Surveiller's laser while taking speed photographs of the ghost of a cedar waxwing" (57). Yet, as in many of Russell's texts, there is considerable ambiguity regarding the phenomenological status of the "ghost birds" in this dystopian future. Not all humans can detect these ghosts; and even those who do claim to have seen them, such as Jasper's ex-wife Yesenia, sometimes come to disavow this. Asked by Jasper whether she recalls the particularly moving song of a ghost nightingale, Yesenia confesses that she always just pretended to detect such revenants in order to comfort her husband. Likewise, a PBS researcher named Suzy comes back from the Monteverde Cloud Forest largely empty-handed in her quest to map the migration of resplendent quetzal ghosts, save for one recording she claims is the song of the extinct violaceous trogon ("twelve down-slurred notes, repeated with a plaintive intensity"). While Jasper admires this trace of "an ancient song forged in Eocene sunlight," Yesenia describes it as being more akin to the sound of "a horny Chihuahua" (59). Here Russell underscores the loss of avian kin who evolved long before *homo sapiens* and how such losses are likely to continue in a world wherein "nobody is [imaginatively] travelling to the future anymore" (58). Surrounded by evidence that biodiversity that has evolved over millions of years is unravelling, many PBS birders respond with eco-melancholic despair that prevents meaningful social reform. Pointedly, Suzy is not around any more to defend her bird recording's authenticity, having committed suicide when she returns from what's left of the rainforest.

Throughout "The Ghost Birds," Russell makes it clear that grief and acute melancholia are key motivators for the paranormal birders, calling into question whether the ghost bird sightings actually exist (as magical realist tropes) within the text's storyworld or whether readers should see these "sightings" as all-too-human projections. Jasper notes that "by the time [he] was born, half of the world's ten thousand bird species of birds had gone extinct" and, while he grows up knowing that this incremental catastrophe is going on in the background, he "never suspected that every bird would disappear in [his] lifetime" (57). Once it's too late to save any remaining species, the shockwaves of this mass extinction reverberate around the world: "In the years following the Great Death, grief-mad humans reported sightings of extinct birds on every continent" (58). In light of such sightings, the reference to paranormal birder Orrine's search for "traces of the ivory-billed woodpecker" (57) is especially significant, given numerous alleged sightings of this bird until it was finally officially declared extinct in 2021. These sightings are explored in a 2009 a documentary titled *Ghost Bird*, a name that is taken by the biologist's doppelganger in VanderMeer's *Area X: The Southern Reach Trilogy*. Though Jasper claims that such sightings were symptomatic of the fact that "the very first paranormal bird-watchers rarely understood what they were seeing and hearing" (58), commonly mistaking ghosts for surviving flesh-and-blood birds, it's clear that eco-melancholia, or a refusal to let go of unspeakable loss, is driving both quests to find some lingering trace of avian presence.

However, "The Ghost Birds" complicates any easy reduction of Jasper and his fellow ghost-birders' obsessions to mere symptoms of acute melancholia. Although he recognizes grief's outsized role in paranormal birding, far more than this is at stake for him in this practice. At various points, he suggests that paranormal birding is a way of holding open a space for transcendent "otherness" and indeterminacy in the face of ecological apocalypse. In her *New Yorker* interview, Russell describes what she sees as an important connection between bird watching and ghost hunting, namely,

> a desire to observe another reality, to experience a vision, a winged visitation. Both are active vigils, underwritten by a special kind of faith. The serious birders I know all have this almost superhuman patience—a really keen and loving attentiveness to the natural world. Wide open ears and eyes.

What might it mean to have such an "attentiveness" in a near-future when the "natural world" has become a specter and extinction is the norm? Jasper gives readers a clue when he says that he hopes to teach his daughter how to identify a ghost bird from its sound alone, and to value "not just the waiting and the listening but the openness to revelation. Which is another way of saying, to being wrong about what is possible and true" (60). In a society where transcendence via technology is the norm (whether via virtual reality, artificial life-support or off world starships), Jasper wants "Starling to stay on Earth with [him]" and worries that "she is losing her dreaming eye—that conjuring eye that is hers alone, the one that can see beyond appearances, into the ultraviolet" (58). For Jasper, what is jeopardized in a world where human beings are no longer sensitive to more-than-human ghosts is nothing less than a "portal" to alternative ways of being. When asked by Starling why she and her father "risk [their] lives to see a bunch of dead birds," Jasper thinks:

> I struggled to formulate a true answer that would not push her farther away from me. I couldn't tell her: You are growing up numb to the universe, numb even to your numbness. You don't know the difference between a screen and a portal. Your eyes cannot distinguish between a digital hallucination and a real ghost. A critical window is closing, Starling. I am trying to hold it open for you, so that you can enter the night.

In associating the ability to perceive ghost birds with an openness to realities beyond those circumscribed within historical culture, Russell's text mobilizes a long symbolic tradition of associating birds with the soul and/or with Romantic transcendence. Yet, rather than merely reflecting Jasper's desire to escape what has become a material hell on Earth, holding open a space for "revelation" has a distinctly secular tinge for him insofar as it is closely related to the capacity for the next generation to imagine alternatives to social injustice and environmental ruin. In this sense, the capacity to be haunted by the ghosts of extinct animals is a crucial pre-requisite to building a better world, one that is a true alternative to petromodernity.

Suggestively, the story's concluding scenes transpire inside the chimney of Chapman Elementary School, a structure famous in Portland, Oregon as the roosting site of the world's largest population of Vaux's swifts. There, thousands of people gather annually on the school's grounds each Fall to watch huge flocks of these birds spiral into their chimney roost in "a miracle of coordination" (61). By 2081, the swifts are extinct and Jasper takes Starling on a quest to view what he hopes will be large flocks of ghosts entering the smokestack one September evening. Beyond its unique history with swifts, Russell's future Chapman Elementary has also become a ruined monument to early twentieth-century ideas of progress, both in its Classical Revival architectural style and its elaborate art, including one New Deal-era marquetry mural by Aimee Gorham. Jasper describes this wood inlaid artwork which frames two doors in some detail:

> two human figures [were] cast as guardians of the portal. A young barefoot girl stood under the tree of life with a dove on one arm, and I swear she looked just like Starling. The wood grain turned an undersea green and mauve as [Starling] spun her light over the doors' engraving: "Send Us Forth to Be Builders of a Better World".
>
> (62)

The mural's motto sums up the mission of a society that believed education was a key to producing future generations capable of improving the world, and its construction by an artist sponsored by the "Works Projects Administration" (WPA) represents an example of such improvement insofar as it reflects humane economic support for distressed citizens during the Great Depression. Likewise, the iconography of the "barefoot girl [standing] under the tree of life with a dove on one arm" could be read as a vision of ecological reconciliation, a cessation of modern western society's war on the natural world and desire to find ways to peacefully co-inhabit the planet with more-than-human others.

But from the vantage point of 2018, it is too late to achieve such harmony: all of the world's birds (and, presumably, many other species) have already become extinct. Russell does not let readers forget that although Chapman school monumentalizes hope for a better future, it was also a structure that rested upon a fossil-fuel foundation, initially heated by a boiler complete with a smoke-stack for its emissions—an archetypal symbol of industrial capitalism and its fatal by-products of greenhouse gases and climate change. Immediately after Jasper and Starling study the WPA mural, they encounter "a stairwell filled with four inches of gray ash" (62) no doubt created by wave after wave of fires that have rendered Portland a dead city. Given the school's complex, seemingly contradictory meanings, readers may be led to ask whether Chapman memorializes unfulfilled hope for social progress or whether it is a monument to anti-ecological hubris. I believe that characterizing it as one or the other obscures a more subtle meaning embodied by this edifice: namely, missed *potential* for a better future based upon the principle of mutual adaptation.

When a large colony of Vaux's swifts took up residence in Chapman Elementary's chimney in the 1980s, a biologist informed the school that continuing to use the old boiler furnace would kill the birds. In order to save the swifts, the school voted to accept the Portland Audubon Society's offer to fund an alternative heating system even though this meant that, until the new infrastructure was operational, pupils would have to bundle up until the birds left on their winter migration. Despite this, students "voted to retire their furnace, piling on parkas and shivering at their desks until the last birds left. The children changed their plumage to save the swifts" (60). This voluntary human adaptation is, of course, a response to the birds' previous adaptation to make their roosts inside of human-built structures once their natural habitat was destroyed; they had been

> forced into the arrangement by humans, who clear-cut the woods and encroached on their homes. When the birds were unable to locate old-growth snags, they adapted to a stone forest of millworks and smokestacks. Later, small bands of humans worked to protect the "chimney corridor."
>
> (60)

On the one hand, the adaptations at work here are asymmetrical insofar as a bird species is "forced" to adapt to anthropogenic habitat loss, whereas human beings appear to have the luxury of *choosing* to accommodate the swifts. However, the text suggests it would be a mistake to see human adaptation to the birds merely as an act of selfless altruism; instead, there is also considerable survival value in choosing to make room for the lives of other species. The choice to fortify the chimney with steel cables in order to protect the swift colony ends up saving the building from a series of catastrophic quakes:

> Chapman Elementary had not been destroyed, and this had everything to do with humans' love of Vaux's swifts. Birds were the reason the chimney still grazed the clouds, a factory-style smokestack with a Dickensian vibe, far better preserved than the ruins of downtown Portland.
>
> (60)

So, in a very tangible way, the school is a monument not only to failed hopes for social progress, but also to a missed opportunity to learn to "build a better world" guided by the principal of co-adaptation as a mutually beneficial survival strategy. As Rob Dunn convincingly argues in *A Natural History of the Future*, the Great Acceleration has been an era marked by anthropogenic extinction of many species, but is also a period in which "an extraordinary diversity of species evolved to depend on humans" (254), whether it be via adaptations to human agriculture or to urban environments. But even as we marvel at the capacity of creatures such as the *Culex pipiens* mosquito to adapt to habitats such as the London Underground railway system, we would do well to recall that "we need to find ways to save the species and ecosystems on which we [humans] depend or might someday depend" (250). Significantly, as a teenager Jasper not only fails to anticipate that all bird species would become extinct in his lifetime; he also did not understand that such losses were harbingers that "*our time would end as well*" [my emphasis] (58). Though what Jasper means by "our time" is somewhat cryptic, I believe he is suggesting that bird die-offs foreshadow the end of the human species, and/or that climate change and mass extinction have led to the loss of a "time" when humans could assume that nature would always be a stable support, that the world would always adapt to our needs.

Jasper's astonishing vision in the story's conclusion reflects his hope against hope that Starling's generation might find a way out of the fossil-fuel-based tunnel vision of her ancestors to "build a better world" via sympoietic co-adaptation with more-than-human others. Disappointed at not sighting ghost swifts entering the chimney at sunset, the two make their way into this abandoned structure, hoping to find traces of the lost colony inside. Instead, they find themselves trapped in the smokestack once a chute access jams behind them, confronted by "a heavy dark where no ghosts roosted, hemmed in by blank brick, out of the top of which we could see the deep-black sky and the rippling light of the stars" (62). Yet what initially seems likely to be their tomb (given that there's no way out except through the top of the chimney) becomes instead a portal to the future. Wounded by a fall after attempting to climb rungs inside the smokestack, Jasper sends Starling up into the dark, marking her smallness "against that epic climb" with great apprehension: "I confess … I did not think, *My daughter is as bright and fleet and brave as a bird. Of course she'll make it out.* I thought something inarticulably sadder" (64). Here, Jasper expresses the paralyzing fear and anticipatory grief with which many parents today regard the future that their children will likely face should actions not be taken to mitigate climate disaster. Russell herself expressed such anxiety in comments about the summer of 2020 when wildfires turned Portland's skies orange: "Our baby's nursery filled with smoke, and those were the worst nights for me, a new kind of claustrophobia and helplessness, a feeling that I could not protect my kids, that every breath was a harrowing event" ("Ecological Crisis").

Yet at a moment when escaping impending doom seems highly unlikely to Jasper, somehow, miraculously, ghost birds begin to materialize out of the darkness to light a way forward:

> I stared up the flue and watched as they illuminated the rungs for Starling, their bodies burning so much more brightly than the dimming bulb of her headlamp. When I looked again, the chimney was shaking apart. Bricks began to lift and dizzy around the cylindrical walls. Blue and gray in the moonlight, course after course of glowing bricks growing wings before my eyes. The bricks were swifts, I realized. … I watched my daughter watching the ghost birds … She began to climb after them. Their light was guiding her out. A held breath of swifts exhaled skyward in a rush, and my daughter was among them, pulling herself onto the school's roof.

As with Jasper's previous ghost-bird sightings, it is entirely possible to read this visitation as wish fulfillment, projections of a mind overburdened with trauma, guilt and grief. Yet a reductively psychological reading obscures the symbolic collaboration at work between Starling and the swifts in this vision. The spectral avians "guide [Starling] out" of the deadly impasse of fossil-fuel capitalism and its failing artificial life-support systems such as her fading headlamp which is "dim" in comparison to the birds' light. Yet as the smokestack (symbolic edifice of phallogocentric petromodernity) collapses, the ghosts of animals rendered extinct by such technology are freed—their spirits rushing skyward like an exhaled "breath" of fresh air. The ghost swifts can be said to haunt such places as reminders of humanity's capacity to destroy nature's ecosystems, but they also signify our capacity to change for the sake of non-human others, holding open the possibility of mutual co-adaptation as a basis for future survival. Although the narrative's final paragraph evokes another reminder of ecological loss in the climate-change era by likening the swifts' blue glow to the "otherworldly sapphire hue you could once see shining through crampon holes in [now extinct] glaciers," the apocalyptic "blackout" that Jasper fears "did not come." Instead, he derives hope from the thought that "so much remains to be seen" if we can attune ourselves more perceptively to why the traces of the dead continue to haunt us: "We have so much more to learn from them. How to pierce the smoke wall of our dulled senses and lift into the unknown. How to navigate the world to come" (58). Karen Russell's future-oriented eco-melancholia anticipates heart-wrenching species loss, but also suggests that it is not too late for early twenty-first-century citizens to find alternative ways to "navigate the world to come" if only we can "learn" to build better worlds in the company of more-than-human others. As she notes in her *New Yorker* interview, it's not enough to draw "dire warnings" of the future in "pointillist detail"; we must also put our imaginations "to new use, and try to conjure a world that does not yet exist, but might. A beautiful future."

Notes

1 The assertion that humanity is causing a sixth major extinction event on the planet analogous to the previous five mass extinctions remains controversial in scientific circles. Although current rates of anthropogenic extinction are well above (up to 100 times higher) the expected "background" rate of non-anthropogenic extinction due to evolution/natural selection, the magnitude of these extinctions has not yet reached the paleontological definition of a "mass extinction" event, defined as a loss of over 75 percent of the world's species within 2 million years or less. Some scientists emphasize that, if current rates of anthropogenic extinction continue, that we will reach "mass extinction" magnitude in a relatively short period of time (relative to earlier mass extinction events), and/or argue that these rates indicate that we are already in the early stages of a sixth mass extinction event despite not yet reaching 75 percent die-offs (see Gerardo Ceballos et al., "Accelerated modern human-induced species losses"). Others are skeptical of such claims. For example, Valenti Rull asserts that it isn't possible to know whether humans will continue to cause high rates of extinction for millions of years into the future, or whether our species will even be around this long, such that our capacity to trigger a "mass extinction" event is not likely (see "Biodiversity crisis or sixth mass extinction?"). Despite such disagreement about whether we are (technically) triggering "mass-extinction"-scale die-offs in non-human species, there is a general consensus in the scientific community that the rise in anthropogenic extinction rates over the past three centuries is alarming and requires intervention to prevent further losses.

2 Here I invoke Lauren Berlant's concept of "cruel optimism" from her ground-breaking 2011 book of that name. I would assert that ecomodernism constitutes a form of "cruel optimism" insofar as it promotes a technocratic vision of the future that upholds problematic modernist ideas of "progress" that are potentially an obstacle to true progress, including a neoliberal vision of the so-called "good life" and denial that over-consumption in the global north is a significant problem. In particular, the fact that this lifestyle has become increasingly unattainable for both citizens of the global north and the global south resonates with Berlant's ideas regarding "cruel optimism."

3 Davis Hucken is repeatedly referred to using his first name (rather than surname) in the novel, so to avoid confusion I will repeat this pattern in my analysis of the text.

4 For studies that illustrate re-wilding's conservationist potential, see Caroline Fraser's *Rewilding the World*, Paul Jepson and Cain Blythe's *Rewilding: The Radical New Science of Ecological Recovery* and Christopher Preston's *Tenacious Beasts.*

5 See Ross Anderson's "Welcome to Pleistocene Park" for an overview of this enterprise.

6 For example, see Naomi Skwarna's interview on the occasion of *Hummingbird Salamander*'s publication, titled "'Activists Are Seen as the Enemy'": An Interview with Jeff VanderMeer."

References

Anderson, Ross. Welcome to Pleistocene Park. *The Atlantic*, April 2017, www.theatantic.com/magazine/archive/2017/04/pleistocene-park/517779/.

Berlant, Lauren. *Cruel Optimism.* Duke University Press, 2011.

Bradley, James. An Ocean and an Instant. *Sydney Review of Books*, August 21, 2018, https://sydneyreviewofbooks.com/essays/an-ocean-and-an-instant.

Bradley, James. *Ghost Species.* Hamish Hamilton, 2020.

Ceballos, Gerardo et al. Accelerated Modern Human-Induced Species Losses: Entering the Sixth Mass Extinction. *Science Advances*, Vol. 1, No. 5, 2015, www.science.org/doi/10.1126/sciadv.1400253.

Christian-Svenning, Jens. Future Megafaunas: A Historical Perspective on the Potential for a Wilder Anthropocene. In *Arts of Living on a Damaged Planet*, edited by Anna Tsing et al. University of Minnesota Press, 2017, pp. G67–86.

Dawson, Ashley. Biocapitalism and De-extinction. In *After Extinction*, edited by Richard Grusin. University of Minnesota Press, 2018, pp. 173–200.

Despret, Viciane. Afterword: It is an Entire World That Has Disappeared. Translated by Matthew Chrulew. In *Extinction Studies: Stories of Time, Death, and Generations*, edited by Deborah Bird Rose et al. Columbia University Press, 2017, pp. 217–22.

Donlan, Josh et al. Re-Wilding North America. *Nature*, Vol. 436, August 17, 2005, pp. 913–14.

Dunn, Rob. *A Natural History of the Future: What the Laws of Biology Tell Us About the Destiny of the Human Species.* Basic Books, 2021.

Economides, Louise and Laura Shackelford, editors. *Surreal Entanglements: Essays on Jeff VanderMeer's Fiction.* Routledge, 2021.

Farrier, David. *Anthropocene Poetics: Deep Time, Sacrifice Zones and Extinction.* University of Minnesota Press, 2019.

Fraser, Caroline. *Rewilding the World: Dispatches from the Conservation Revolution.* Picador, 2010.

Ghost Bird. Directed by Scott Crocker, Small Change Productions, 2009.

Haraway, Donna. *Staying with the Trouble: Making Kin in the Chthulucene.* Duke University Press, 2016.

Heise, Ursula K. *Imagining Extinction: The Cultural Meanings of Endangered Species.* University of Chicago Press, 2016.

Jepson, Paul and Cain Blythe. *Rewilding: The Radical New Science of Ecological Recovery.* Icon Books, 2020.

Marris, Emma. *Rambunctious Garden: Saving Nature in a Post-Wild World.* Bloomsbury, 2011.

Mortimer-Sandilands, Catriona. Melancholy Nature, Queer Ecologies. In *Queer Ecologies: Sex, Nature, Politics, Desire*, edited by Catriona Mortimer-Sandilands and Bruce Erickson, Indiana University Press, 2010, pp. 331–58.

Preston, Christopher. *Tenacious Beasts: Wildlife Recoveries That Change How We Think about Animals.* MIT Press, 2024.

Rull, Valentí. Biodiversity Crisis or Sixth Mass Extinction? *EMBO Reports*, Vol. 23, 2022, n.p.

Russell, Karen. The Ghost Birds. *The New Yorker*, October 11, 2021, pp. 56–64.

Russell, Karen. Karen Russell on Home, Monsters, and Magical Realism. Interviewed by Bradley Sides, *Chicago Review of Books*, July 22, 2019, https://chireviewofbooks.com/2019/07/22/karen-russell-on-home-monsters-and-magical-realism/.

Russell, Karen. Karen Russell on Ecological Crisis. Interviewed by William Davidson, *The New Yorker*, October 4, 2021, www.newyorker.com/books/this-week-in-fiction/karen-russell-10-11-21.

Russell, Karen. *Orange World and Other Stories.* Knopf, 2019.

Russell, Karen. *Swamplandia!* Random House, 2011.

Sperling, Alison. "Second Skins": A Body-Ecology of Jeff VanderMeer's The Southern Reach Trilogy. *Paradoxa*, Vol. 28, Special issue: Global Weirding, 2016, pp. 214–39.

van Dooren, Thom. *Flight Ways: Life and Loss at the Edge of Extinction.* Columbia University Press, 2014.

VanderMeer, Jeff. "Activists Are Seen as the Enemy": An Interview with Jeff VanderMeer. Interviewed by Naomi Skwarna, *Hazlitt*, May 19, 2021, https://hazlitt.net/feature/activists-are-seen-enemy-interview-jeff-vandermeer.

VanderMeer, Jeff. *Area X: The Southern Reach Trilogy.* FGS Originals, 2014.

VanderMeer, Jeff. Hauntings in the Anthropocene. *Environmental Critique*, Vol. 3, No. 7, July 17, 2016, https://environmentalcritique.wordpress.com/2016/07/07/hauntings-in-the-anthropocene.

VanderMeer, Jeff. *Hummingbird Salamander.* Farrar, Straus and Giroux, 2021.

4 Reproductive Rights in the Context of Deep Time

In the *longue durée* of our species' history on Earth, the last 200 years have been a period of remarkable change. An online video for the American Museum of Natural History (AMNH) graphically illustrates the trajectory of humanity's expansion in time and space, from its humble beginnings in Africa to its current status as a global geological-scale force, foregrounding the following astonishing fact: "It took most of human history for our population to reach 1 billion—and just over 200 years to reach 8 billion." In November of 2022, the world's human population reached 8 billion, and according to 2022 United Nations projections, will continue to grow (albeit at a slower rate) to a projected 10.4 billion before levelling-off around 2100.[1] The UN's "Global Issues: Population" webpage features an equally striking video presentation titled "Eight Billion Lives, Infinite Possibilities: The Case for Rights and Choices," which argues that concerns about world population growth should be re-framed as an issue of women's rights. Instead of asking whether there are "too many" or "too few" people, the question should be "whether women are able to make their own reproductive choices," which is not the case for 44 percent of women globally who are not allowed "to make their own decisions about their reproductive health, contraception and whether to have sex." Taken together, these testimonials encapsulate an "anthropos"-level understanding of world population growth in the twentieth and twenty-first centuries as a major challenge to ecological sustainability on the one hand, and a "homo" level framing of the same phenomenon as a by-product of social injustice on the other. The visuals in these divergent video presentations also reflect the different scalar perspectives entailed in both kinds of analysis: the AMNH's film adopting a "god's eye" view wherein pointillist dots representing human population on a global map multiply gradually then rapidly expand over long timescales, whereas the UN video depicts images of human-scale women (and children) of different ethnicities that morph/merge through different sequences that correspond to the narration. Although often framed as an "either/or" issue, this chapter will make a case for the importance of drawing upon both "anthropos"- and "homo"-scale framing in order to forge a novel reproductive politics that is truly committed to multispecies justice. As with the related issues of decreased biodiversity in non-human species explored in the last chapter, human population demographics stand to have long-term consequences in the Anthropocene era, so the stories we tell about these trends and reproductive rights matter.

DOI: 10.4324/9781003243946-5

Given Anthropocene theory's focus on global socio-environmental change, it should not be surprising that there has been renewed scholarly discussion around population growth as a challenge that needs to be addressed in order to insure a biologically diverse future: one in which both human and more-than-human life can flourish. Geometric increases in human population in the last two centuries, combined with the current extinction crisis in wild animal populations documented in the last chapter, suggest that it is not entirely "wrong" (contra the UN website video) to ask questions about how more much human population growth is sustainable if we are concerned about maintaining habitat for viable populations of more-than-human lifeforms. Hence, major theorists of the Anthropocene, such as Timothy Clark, Donna Haraway, Hannes Bergthaller and Eva Horn have all addressed population pressures,[2] although controversy remains around the question of how to frame this issue in a constructive manner that honors human rights and avoids the pitfalls of politically reactionary formulations while not discounting the enormous impact our expanding cumulative footprint has had on earth systems. As documented in David Attenborough's *A Life on Our Planet*, as of 2020 half of the Earth's fertile land was being used for agricultural production, half the world's rainforests were gone, freshwater wild animal populations had been reduced by over 80 percent, and 30 percent of oceanic fish stocks had been reduced to critical levels (95–7). While capitalist over-consumption (of things like meat and fossil-fuel-powered energy) is undeniably a major factor driving such statistics—especially in the global north where per capita consumption is much larger than it is in the global south—this does not mean that global population growth is an entirely negligible issue that we can assume will "take care of itself" in the future without significant social progress being made in arenas such as female/trans-empowerment.

Indeed, as Ursula Heise points out in *Sense of Place, Sense of Planet*, as of the 1990s "issues of women's rights, access to education, and reproductive health … frequently [occupied] center stage in discussions of global demographic trends" in population. For good reason: projections for a population plateau by the end of the twenty-first century rely upon factors such as a continued reduction in fertility rates (i.e. the average number of children born to each woman globally) that has been slowing since the mid-twentieth century, a trend largely driven by global improvements in women's rights, including the access to education, employment and healthcare (including reproductive services). Yet in the years following Donald Trump's presidential election in 2016 and 2024, it is precisely women's reproductive autonomy, along with access to education that addresses gender, sexuality and structural inequity, that have come under fire in the United States—a reactionary trend mirrored in non-democratic and far-right regimes in both the global north and south. The US Supreme Court's overturning of Roe v. Wade in 2022 ended women's access to abortion as a constitutionally guaranteed right. Following the November 2024 elections abortion is still banned at contraception or after six weeks of gestation in 16 states, despite the fact that in 7 out of 10 states where this issue was on the ballot, voters decided in favor of individuals' rights to make their own reproductive decisions. Meanwhile, far-right activists are seeking a national ban on medical abortions and to eliminate gender/sexuality studies (and critical race theory) in elementary, secondary and higher education.

As the Center for Reproductive Rights documents, although significant progress has been made globally over the past 25 years towards recognizing women's reproductive rights (including the "liberalization of abortion laws" in over 60 countries), as of 2024, 21 countries still completely prohibited abortion (affecting 6 percent or 111 million women of reproductive age) and 40 percent of women still "live under restrictive laws," defined as an "inability to access safe and legal abortion care." Ideologically, there is often a strong correlation between inegalitarian politics and the denial of women's reproductive rights. For example, a 2023 PerryUndem National U.S. Survey titled "Assessing the State of Public Opinion Towards Women, Gender, Equality—and Abortion" found that "sexist beliefs are very strongly correlated with views towards abortion," with most anti-abortion Republicans holding such views (54–64). During President Bolsonaro's time in office in Brazil, lawmakers introduced at least 30 bills to tighten abortion restrictions and introduced legislation intended to dissuade rape survivors from seeking abortions by requiring healthcare providers to report such women to the police. Fundamentalist regimes in countries such as Iran have passed reproductive laws such as the 2021 "rejuvenation and support of the family" bill that restricts women's access to contraception and education about reproductive health, while supporting programs "encouraging women to have children and denouncing decisions to remain single, have fewer children, or have abortions" ("Iran"). Likewise, since the Taliban takeover of Afghanistan in 2021, women's and girls' rights have reverted to pre-2002 levels, effectively erasing progress on women's rights in the intervening 20 years, including the right to "education, work, health, access to justice and freedom of movement, attire and behavior" ("As Women Worldwide"). In sum, despite global improvements in many nations, regressive trends reveal that women's human rights (including reproductive rights) are *not* guaranteed, which suggests that projections for future population stabilization are likewise contingent. Indeed, as our climate-altered world becomes more chaotic—subject to unprecedented droughts, destructive weather events and increasingly frequent disease outbreaks or pandemics—the temptation to embrace fundamentalist or totalitarian social order is likely to become more appealing, particularly in the absence of effective environmental policy to mitigate against these threats. Hence, as health science professor Céline Delacroix notes in "Ecology and Empowerment," there is a great need to demonstrate ways that "population growth can be effectively addressed by expanding autonomy and the exercise of reproductive rights for women and by removing barriers to safe and effective family planning," a connection that she argues is "not sufficiently known or understood" or is even "systematically avoided in key international policy instruments."

Within the arenas of environmental philosophy and ecocriticism, the synergy between population pressure and the denial of rights for women has long been understood by ecofeminists and environmental justice activists. Today's scholars of the Anthropocene who express concern over population would do well to recall debates about this issue between deep ecologists and ecofeminists in the 1980s and 1990s, particularly with regard to how to frame these concerns. In *Deep Ecology* (1985), Bill Devall and George Sessions attempted to distinguish their biocentric ecological platform from "reformist" approaches, addressing what they felt were the root causes of environmental problems such as anthropocentrism, excessive consumerism and overpopulation. One of deep ecology's "basic principles" is that "the flourishing of human life and cultures is compatible with a substantial decrease of the human population" and that "the flourishing of nonhuman life requires such a decrease" (70). Though Devall, Sessions and the Norwegian philosopher Arne Naess are vague as to what constitutes a "substantial decrease" in global population and how this goal would be achieved, EarthFirst! co-founder, deep ecologist and self-proclaimed "eco-warrior" Dave Foreman announced in his book *Man Swarm* that the global population must be brought down to 2 billion people in order to preserve the world's remaining "wild things" and focused on the problem posed by immigrant populations as a driver of overpopulation and threat to wilderness in the United States. Foreman's xenophobia and dehumanization of peoples of color, the poor and migrants in numerous public statements underscores the risk for legitimate concerns over the impacts of population pressures to be co-opted into regressive, racist and patriarchal ideologies that reinforce a colonialist mindset. Ecofeminists Greta Gaard and Lori Gruen address this threat head on in "Ecofeminism: Toward Global Justice and Planetary Health" (first published in 1993). Although they acknowledge that "global population problems are real and a key factor in environmental deforestation, desertification, and soil erosion," they are also critical of "top-down" policies that mandate "population control" in the developing world while ignoring factors such as women's rights, wealth inequity and the developed world's disproportionate consumption of resources, noting that "to blame environmental degradation and economic distress on the population of the Third World is just another form of racism, sexism and victim-blaming" (165). Taking issue with a focus on "overpopulation" evoked in holistic philosophies such as deep ecology that suggest that "the problem is one of mere numbers," they advocate for a more fine-grained and historically informed approach:

> From an ecofeminist perspective … the so-called problem of overpopulation may be more adequately named and addressed by looking at the intersection of several factors: reproductive control, socio-economic conditions, the material status of women, and … the (unjust) distribution and consumption of 'natural resources.' The first, reproductive control, includes access to free, safe methods of contraception, including elective abortion, it includes freedom from compulsory sterilization as well as compulsory motherhood; it includes the freedom to control the number and spacing of one's children … but birth control programs alone have been shown to account for only 15 to 20% of the overall decline in fertility rates; the remaining percentage is attributed to socio-economic factors such as economic security, increased literacy rates, better education, better health care, and better job opportunities for women. (166)

In addition to illuminating links between population pressure and social injustice, ecofeminist analyses like Gaard and Gruen's also underscore the inadequacies of top-down technocratic "solutions" to such global challenges if such policies ignore equity issues. As these authors observe,

> When overpopulation is seen as the main cause of the current environmental crisis, women are often blamed for this burgeoning population. On this view, the solution to overpopulation is the development and use of biotechnologies by which women's bodies and procreative powers are controlled. Medically dangerous contraceptive methods such as the IUD (intrauterine device), use of the drugs Depo-Provera and Norplant, or compulsory sterilization are offered to poor women as the responsible alternative to bearing children that have a small chance of survival.
>
> (157)

Coercive attempts to persuade women in developing nations to use medically risky technologies (often as guinea pigs for big pharma) are an example of the perils of neocolonialist technocracy. But even in the case of medically safe contraceptives, it is naïve to think that the mere existence of such technology will (in itself) solve population challenges. The complexity of factors that influence reproductive decisions—including economic, cultural, racial, historical and religious disparities among different groups of women *and* different challenges faced by non-binary or trans persons—demands equally complex, nuanced analysis and policy-making.

Anthropocene theory adds a new layer of complexity to population debates insofar as contemporary humanity's geological-scale environmental impacts and the duration of such changes over deep time pose potential challenges to previous ways of framing such controversies. Timothy Clark addresses these challenges in *Ecocriticism on the Edge*, including how "scalar effects" complicate social justice frameworks. For him, global population pressure has become a "hyperobject" in Morton's sense of being a phenomenon whose massive impacts are "global and non-localizable" (87) and whose effects must also be understood intergenerationally over long periods of time. Clark implies that, as a scalar effect, individual decisions about whether or not to have a child are analogous to an individual's choices regarding whether or not to drive a car, or to fly long distances, etc., because while these decisions are ecologically trivial on an individual level (and can even be life-enhancing), when magnified on the order of billions of people making similar choices the cumulative environmental impacts can be ecologically destructive. As with other such challenges, Clark sees "a growing population" as "harbour[ing] emergent properties" that open up a "latent rift" between an increasingly "evasive" politics that valorizes individual choice and a "narrowly populationist agenda [that] risks endorsing present or immediate injustice" (83). Hence, for Clark Anthropocene framing "overpopulation" becomes another example of an "insidious … clash between the human rights agenda and the pressures of environmental integrity and safety" (86). As an example of this, he cites Leon Kolankiewicz's studies of traditional slash-and-burn agriculture in Honduran rainforests which show that such practices are "*sustainable when population density is low, but destructive when it is too high*," concluding that ("tragically") the cumulative effects of "rightful claims to subsistence agriculture" can—if practiced on a large scale—erode rainforests just "as well as the large-scale logging companies that do so far more culpably" (84, emphasis in original).

Likewise, Clark is critical of a shift in the 1990s away from considering the demographics of population growth to what he feels is an exclusive focus on women's rights. As evidence of this, he discusses a shift in the years following the International Conference on Population and Development (ICPD) held in Cairo in 1994, whereby consideration of "geomorphic, large-scale effects of overpopulation" gives way to a "more narrowly conceived social justice agenda" that emphasized

> the closely related yet crucially distinct issue of improving the autonomy of women, understood to be the most decisive issue affecting pregnancies. A priority becomes helping all girls to have access to education, and women to family planning, with contraception made easily available.
>
> (82)

Asking "who would take issue with such obvious goods?" Clark acknowledges that in practice the interests of "populationists" and "social justice" advocates should align on the issue of women's empowerment. However, citing Don Weeden and Charmayne Palomba's work, he claims that these interests do not (in fact) fully overlap because in rights discourse women's immediate needs in the present time trump "the longer time-scale effects (with their massive scenarios of human and nonhuman misery)," which are "overlooked or assumed to take care of themselves" (83). As an example of this aporia, Clark cites social justice advocates' critique of "population control" measures that "violate the rights of women" (as was the case in Gaard and Gruen's critique of dangerous forms of contraception). This implies that Clark is critical of a bias towards protecting individual rights that ignores the long-term *collective* consequences (for humans and more-than-human beings) of refusing to adopt "population control" methods.

For all its provocative insight and power to re-open important debates about the future of population pressure, Clark's account of "overpopulation"—much like the generalizations of previous deep ecologists—has a tendency to elide important distinctions made by social justice advocates, which is ironic given the debt his methodology owes to deconstructionism. For example, Clark worries that, given dramatic world population growth, aspirations to make "Western-like levels of prosperity and consumption" more "widespread" constitutes "an evasive and anthropocentric agenda whose implementation would actually see accelerated levels of habitat loss across the world to support levels of human consumption set to exceed capacity after the short term" (81). Thus, he concludes that "the seemingly obvious and just policy of calling for a massive redistribution of wealth from North to South" would entail "a catastrophic increase in energy consumption" in order to achieve levels of development necessary to "reach a stable or even declining population" in developing nations. Yet most social justice theorists do not call for neoliberal capitalist over-consumption and waste to become universal; in fact, either implicitly or explicitly, the "greater economic equity" such thinkers often advocate entails a significant *decrease* in per capita western levels of consumption even as living standards in developing nations would increase to better meet basic needs for clean water, food, housing and education. Likewise, it is problematic to suggest that, in advocating for women's rights to resist unhealthy "population control" methods, feminists and social justice representatives "became strange bedfellows of the Vatican and other conservative religious advocates" (81). There is a false equivalency here insofar as religious conservatives rarely represent themselves as champions of women's rights or reproductive freedom; instead, the "rights of the unborn" almost always eclipse any concern for the rights of mothers. In asking "who would [possibly] take issue with such obvious goods" as helping girls to have access to education, easily available contraception and reproductive autonomy, Clark ignores the very considerable (and growing) *resistance* to these things on the part of fundamentalist and far-right political movements documented at the outset of this chapter. Indeed, his account of the Anthropocene illustrates ways theorists working in this area could learn a lot from previous ecofeminist philosophy.

As discussed in my "Introduction," one of Clark's more compelling claims in *Ecocriticism on the Edge* is that the scalar effects he analyzes are extremely difficult to capture in conventional literary forms. Hyperobjects such as climate change and overpopulation are global and non-localizable, not reducible to humanist-scale art which focuses on individual people caught up in dramas that typically have a short duration (affecting one or two generations at most). While it is certainly true that the classic bourgeois novel is limited in its capacity to represent "humanity en masse," speculative fiction has undertaken avant-garde thought experiments that raise important questions about women's reproductive rights in the context of environmental concerns. In *The Handmaid's Tale* and *The Testaments*, Margaret Atwood illuminates how patriarchal violence, religiously sanctioned control over women's reproduction and the exploitation of feminized "nature" constitute self-reinforcing feedback loops with deleterious social and environmental consequences. The novel's insight into how different forms of nature in Gilead (pastoral gardens decorating Commanders' homes, offshore plantations and toxic sacrifice zones) reflect sexist, homophobic, racist and colonialist politics registers an important critique of "homo"-level social and environmental injustice. However, despite these merits, these novels have little to say about *multispecies* justice and do not consider politics within the context of deep time, except via fleeting references to women's "biological destiny" which may hint at latent evolutionary essentialism underwriting Gilead's theocracy. It is therefore illuminating to contrast Atwood's work with Anthropocene-era texts that bring "anthropos"-level deep time into conversation with "homo"-level historical time. Louise Erdrich's *Future Home of the Living God* expands upon Atwood's ecofeminist insights to explore particular challenges that women of color face when denied reproductive rights, excavating the roots of such oppression in colonialism and its afterlives but also (crucially) deconstructing ways that white supremacist reproductive politics are mobilized by social Darwinism and this ethos's racist, sexist and speciesist understanding of deep evolutionary time. I will argue that Erdrich's novel updates Gilead for the neoliberal and climate-change era, and in doing so exposes connections between gender, race and species-based ideology that often goes hand in hand with attempts to deny women's reproductive rights.

Gilead's (In)human Geography: of Narrow Rooms, Walking Wombs and Colonial Tombs

Margaret Atwood's *The Handmaid's Tale* is paradigmatically a novel of confinement. When we first meet the anonymous female narrator, she is among a group of women being held in a gymnasium, awaiting military re-assignment to a more permanent place of detention—a claustrophobic room in Gilead's domestic sphere that she occupies as a Handmaid but, significantly, refuses to call her own (8). Like "Jane" in Charlotte Perkins-Gilman's "The Yellow Wallpaper," Offred/"Jane" narrates much of her tale from this confined space, a room in which her life has been narrowed to fit the demands of motherhood under a patriarchal regime. Time in this room has no meaning, being reduced to cycles of endless waiting punctuated by repugnant routine, an absence the Janes fill via imaginative observation and projection: is that wreathed "blank space" (Atwood 7) on the ceiling a place where a chandelier once hung? Why are there "rings and things" (Gilman 303) on the walls and bars on the windows? Anonymous rooms haunted by the traces of previous women who tried to resist (tearing wallpaper, scratching messages into closet walls) but who could ultimately only attain freedom via psychic or physical death.

Given the stifling confinements endured by women in Gilead—from their color-coded, restrictive uniforms to their circumscribed ability to travel outside the home to their subjection to constant surveillance—there is a strange exhilaration in Offred's dream of attempting to escape with her daughter through woods on the US–Canada border. Running through "bracken" and "uneven ground" strewn with rocks and dead branches, she can "smell the damp earth" (74) and anticipates plunging into cold water, but instead hits the ground to avoid gunshots. Here in the wilderness, at the height of trauma curled up over her child, she suddenly feels

> calm and floating, as if I'm no longer in my body; close to my eyes there's a leaf, red, turned early, I can see every bright vein. It's the most beautiful thing I've ever seen ... There's breath and the knocking of my heart, like pounding, at the door of a house at night, where you thought you would be safe.
>
> (75)

A moment of expansive wonder and yearning for freedom before she's taken to a sunken place that's metonymic for her narrow room:

> it's too late, we come apart, my arms are held, and the edges go dark and nothing is left but a little window, a very little window, like the wrong end of a telescope … through this window I can see, small but very clear, I can see her, going away from me, through the trees which are already turning red and yellow, holding out her arms to me, being carried away.
>
> (75)

For Offred, this is the "worst" kind of dream she has as a Handmaid, because in it she not only relives the trauma of being separated from her daughter but also the experience of being estranged from her own former self, observing that self from afar as she's reduced to existence as a "two-legged womb" (136) in the service of a totalitarian patriarchy. As Linda Kauffman observes in her essay "Special Delivery: Twenty-first Century Epistolary in *The Handmaid's Tale*," in each of the novel's locales, "Offred situates herself in a landscape that is simultaneously psychic, physical and political" (222). Building upon work by Deborah Hooker, Lauren A. Rule, Christabelle Sethna and Julia Kuznetski, I would add "ecological" to Kauffman's list, and will explore parallels between Gilead's attempts to dominate women's reproduction and its catastrophic environmental politics. Control of women's and nature's fertility is central to Gilead's future viability, yet, ironically, its militaristic policies have put both in jeopardy, the creation of toxic environmental "sacrifice zones" having become so widespread that fertility rates have fallen dramatically. I will examine the novel through an ecofeminist and environmental justice perspective, focusing on how Gilead's patriarchal and white supremacist regime creates rigid hierarchies and a eugenicist agenda as responses to socio-ecological crisis. Although *falling* rather than rising population is the central pressure that mobilizes this autocracy, the novel is nevertheless important as a warning about how tenuous women's rights (and those of other marginalized groups) could become in the face of growing ecological instability and the allure of politically reactionary "solutions" to growing chaos.

At the outset, it is important to specify that Atwood has repeatedly emphasized that her *Handmaid* books are not sweeping condemnations of Christianity per se, but rather studies in how totalitarian regimes can (and have) historically cloaked themselves in a theocratic guise in order to justify inhumane political policies. Although some libraries in the US have recently banned these books on the grounds that they are "sexually explicit" and "anti-Christian,"[3] careful consideration of Gilead's ideology reveals the extent to which the regime's Christian fundamentalism may mask an underlying pseudo-scientific Darwinian ethos that justifies oppressive policies by evoking "Nature's norm" (*Handmaid's Tale* 220). According to this ideology, autocratic white male leaders (the Commanders) have both a God-given and *biological* mandate to restore "natural order" to previously degenerate western societies. For example, when Offred's Commander celebrates the regime's establishment of a society in which women can "fulfil their biological destiny in peace" as one of its most important accomplishments, this suggests that something more than casual chauvinism guides his philosophy. I will have more to say about the history of social Darwinist ideology in my analysis of Erdrich's *Future Home of the Living God* as I believe this text is more explicit in deconstructing this ethos than *The Handmaid's Tale*. Nevertheless, it is suggestive that although Gilead purports to be a theocracy organized around fundamentalist Christian faith its leaders privately endorse a biological essentialism that's very much in keeping with a social Darwinist ethos insofar as it shapes public policies such as racial segregation, strict policing of heteronormative sexuality and a denial of women's rights. As Paul Moffett usefully highlights, it is not merely infertility that motivates Gilead but more specifically concern over "plummeting Caucasian birth rates" (as Professor Pieixoto puts it in the "Historical Notes" section) relative to "non-Caucasians." Hence, the regime puts into place what is, essentially, a eugenics program whereby fertile white women are forcibly conscripted into becoming baby-making machines while women of color are removed to "colonies" outside of Gilead's boundaries. As we shall see, this society segregates both people and landscapes into separate spheres whose sole purpose is to benefit its white supremacist, male elite.

In both *The Handmaid's Tale* and *The Testaments*, Gilead's gendered and racialized geography not only demarcates place but is also informs/deforms the bodies of its "othered" citizens. The regime's organization of space is essentially patriarchal and colonial, comprised of: the metropole (divided between gendered public and private spheres); colonies (either agricultural or toxic wastelands); and wilderness or wild lands. Each of these domains is conceptually linked with Gilead's patriarchal and puritanical attitudes towards sexuality. In the metropole, the decorative gardens that surround the homes of elite Commanders are a controlled, pastoral space that's an analogue of their Wives' largely aesthetic function. Although not all Wives are infertile, the wide use of Handmaids as surrogates in Gilead's elite suggest that many may be, along with their husbands (though men are never blamed for infertility). Although Deborah Hooker's complex interpretation of floral imagery in *The Handmaid's Tale* explores a "subversive" and pagan excess in Offred's description of Serena Joy's "swoon"-inducing garden, she also acknowledges that "flowers are still allowed" in the regime's domestic sphere because "Gilead can depend on the way monotheism has fundamentally 'renatured' flowers to serve its own particular ends" (287). Just as women's fertility is allowable only if it is tightly controlled by the regime, decorative gardens are permitted provided their fecundity is subjected to harsh management. Symbolically, the Commander's Wife Serena Joy serves as an emissary of the regime in this regard when she violently removes seedpods from plants in her garden with a pair of shears, committing a "blitzkrieg … on the swelling genitalia of the flowers" (153). Serena Joy's over-zealous pruning also reflects her resentment that because she has been unable to produce children her household must take on a Handmaid, reducing her own status to a decorative spouse whose "power" is wholly dependent on her husband's whims. As is the case with Commander Judd in *The Testaments*, she may be discarded for another wife as easily as a gardener might decide to plant a new bed of annuals.

In contrast to the urban elite's showy decorative gardens, forms of fertility that are more necessary for human survival such as agriculture and the biologically necessary Handmaids are rendered as *invisible* as possible in Gilead even as they are ruthlessly exploited by the patriarchy. Food crops such as "tomatoes" and those vital for clothing such as "cotton" (248) are grown in colonies that are geographically distant from the fatherland, just as fertile Handmaids are (for the most part) sequestered in rooms that render them invisible to Wives/Commanders that occupy the central rooms in large households. Women's clothing likewise reflects a misogynistic dualism whereby relatively privileged Wives wear blue, symbolizing "purity" analogous to the Virgin Mary (often depicted wearing a blue veil in Christian iconography) whereas Handmaids wear red robes signifying their "whorish" (115) and debased sexual servitude. Handmaids are also forced to wear white bonnets that serve the dual purpose of making their faces invisible to onlookers (emphasizing their anonymity) and blinkering their own perspectives. Indeed, one of the most chilling consequences of Offred's conditioning is her internalization of Gilead's misogynistic panopticon. She vacillates between feeling her identity has been reduced to a womb, having "congealed around a central object, the shape of a pear, which is hard and more real than I am" (73–4) and not wanting to see her own body because she doesn't want "to look at something that determines [her] so completely" (63). Offred's life-giving fertility is converted into a metonymic womb controlled and exploited by others; specularly, this organ comes to appear "more real" (valued) than she is as an individual, but is also something that the regime stigmatizes as obscene, "shameful" (63) and properly hidden.

Likewise, Gilead not only restricts the cultivation of life-giving food to colonies (whose geographical distance makes them invisible from the metropole) but has also created a non-white slave class to perform agricultural labor. Early on in the novel, Offred watches an official news report on television which focuses on Gilead's bombing of Detroit and "resettlement of the Children of Ham" to "National Homelands" (colonies) in places such as North Dakota, where they are "supposed [to] farm" (84). Historically, people of color were associated with the biblical Ham's "cursed" descendants, a stigma that was often used to justify slavery. Like Handmaids forced to bear children for the elite but not allowed to enjoy the fruits of their labor, people of color are relocated from urban to rural areas and (implicitly) forced to work on "cotton" and fruit plantations that echo the history of slavery in the United States. Gilead's neocolonial and racist biopolitics serve multiple purposes: outsourcing agricultural production, greatly reducing labor costs and segregating the metropole, creating an exclusively white population in the center of power. Critics such as Christabelle Sethna have linked Gilead's racial segregation to its reproductive agenda, describing its republic as "a white supremacist, patriarchal, evangelical Christian, theocratic stronghold bent on improving the white birth rate" (1). Sethna illustrates how this reading is consistent with Atwood's concerns about the rise of the New Christian Right (NCR) in the 1980s, particularly its anti-abortionist stance as a reactionary response to "the decline in the Caucasian birth rate in North America and northern Europe" (quoted in Howells 96). Thus, Gilead mobilizes both racist and misogynistic ideology in making abortion a crime punishable by death and likewise attempts to dominate the natural world by dividing landscapes into aestheticized gardens, fertile (but distant) agricultural land worked by slaves and toxic, sacrifice zones populated by those deemed to be human refuse.

As Offred's rebellious friend Moira relates, the regime sends its "discards"—such as Handmaids who don't conceive after serving in three households, older infertile women, political dissidents and "gender traitors"—to non-agricultural "colonies" where they are expected to "spend their time cleaning up" military and environmental disasters. In war-torn locales, this entails burning the bodies of fallen soldiers or plague victims; in other zones, laborers are forced to manage "toxic dumps and radioactive spills" (248). Because workers in these toxic zones are not given protective clothing or much food, life expectancy is reduced to "three years maximum," so being sent to such a colony is (effectively) a death sentence. Significantly, all "unpersons" sent to toxic colonies are degraded to the status of unwanted, feminized "others," as is reflected in the fact that all those sent to these zones are forced to wear long, gray dresses. Implicitly, this is also a place where non-binary LGBTQ people are also sent, as "gender traitors." As in the agricultural colonies, there are important parallels between enslavement, reproductive/sexual politics and the use/abuse of land in the sacrifice zones. The toxic colonies are ruined, infertile wastelands created by Gilead's war machine, which produces "chemical- and biological-warfare stockpiles and toxic waste" (304), all of which cause infertility in those sent to such areas if they are not already this way before being deported.

It is no accident that women and homosexuals are most commonly the victims of this deportation. The threat of deportation to environmentally ruined areas hangs over the heads of all women who refuse to allow their bodies to be controlled by the state and to all LGBTQ peoples who reject cisgendered identity—in short, to citizens whose "infertility" from the perspective of heteronormative patriarchy leads them to be categorized as "ruined" and disposable. Gilead's leaders believe they have the power to determine which areas of the Earth and which human populations can exploited and which can be sacrificed. As Julia Kuznetski notes, *The Testaments* updates the scope of Gilead's ruination to include extractive industries, as when Offred/Jane's teenage daughter Agnes (who has been adopted by an elite Commander's family) muses about the liability that women's embodiment poses under the regime:

> The adult female body was one big booby trap as far as [she] could tell. If there was a hole, something was bound to be shoved into it and something else was bound to come out, and that went for any kind of hole: a hole in a wall, a hole in a mountain, a hole in the ground. There were so many things that could be done to it or go wrong with it … that [she] was left feeling [she] would be better off without it.
>
> (83)

Although Kuznetski does not speculate about the specific modes of extraction that are reflected in Agnes's phobic musings (noting only that her fears conflate "corporeal" and "environmental properties" (294)), both *The Testaments* and *The Handmaid's Tale* suggest that Gilead is in the business of fossil-fuel extraction. Although "not many things are plastic, anymore" (*Handmaid's Tale* 27) under this autocracy and gas was already "expensive and in short supply" (85) when the regime took over, the elite are still chauffeured in private cars and there are service vehicles for specific purposes (such as "Birthmobiles"). Indeed, *The Testaments* makes it clear that Gilead not only uses fossil fuels but also denies the existence of Climate Change, sparking protests among Canadian activists (including Jane's second daughter Nicole) who carry signs with slogans such as: "GILEAD, CLIMATE CHANGE DE-LIAR! GILEAD WANTS US TO FRY!" and which depict "forest fires and dead birds and fish and people" (48).

Wilderness or wildlands constitute a third sphere which delimits the reach of Gilead's power. "Wooded Hills" such as the "Appalachian Highlands" demarcate the "front lines" of its military expansion, places where the regime's attempts to eradicate rival religious factions are met with resistance from groups such as "Baptist guerillas" (368). Wildlands are also places where Quakers and the Mayday resistance movement help Handmaids and other defectors to flee from Gilead. The forests and rugged coasts of Maine become routes for Mayday's "Underground Femaleroad," recalling the Underground Railroad for escaped slaves during the US Civil War. Offred/Jane and her daughters all use this network to escape, which suggests that wilderness in *The Handmaid's Tale* and *The Testaments* is a space of freedom along the lines that Andreas Malm analyzes in his essay "In Wildness is the Liberation of the World: On Maroon Ecology and Partisan Nature." Gilead is a colonizing power that seeks to dominate subjugated classes of human beings *and* is hostile towards uncultivated nature. It is into such unconquered spaces that refugees in Atwood's novels flee, mirroring the quest of prior exploited peoples seeking freedom under colonialist capitalism (such as the Maroons and Jewish partisans who Malm discusses). Although there are few descriptions of wild places in either *The Handmaid's Tale* or *The Testaments*, passages that do depict wilderness (such as the one that I cite at the beginning of my analysis of *The Handmaid's Tale*) are notable for their expansive, transcendent quality relative to Gilead's restrictive jurisdiction. For example, when Agnes and her half-sister Nicole float under cover of darkness on the Penobscot River in Maine's wilderness on their way to freedom in Canada and see a crescent moon reflected in the water, Agnes exclaims, "Look … I've never seen anything so beautiful! It's like a trail of light!" (*Testaments* 367–8) echoing her mother Jane's wonder at an autumnal leaf ("it's the most beautiful thing I've ever seen") that she recalls in dreams about her abortive attempt to escape through forests at the US–Canadian border. In sum, as spaces "outside" the reach of Gilead's totalitarian geography wildlands are a symbolic threat to its authority and, as such, are spaces of conflict where the regime attempts to reassert its control but (more often than not) is thwarted. In *The Testaments*, Aunt Lydia reveals that Gilead has "an embarrassingly high emigration" rate as more and more Handmaids (and their offspring) attempt escape through wilderness "buffer zones [that] are too permeable" (112). She confesses that "the wilder patches of Maine and Vermont are a liminal space not fully controlled by us, where the natives are … prone to heresies," a problem that the zealous Aunt Elizabeth suggests could be remedied by "clear-cutting [ing] the entire area," a proposal that is supported by powerful men like Commander Judd (113). From the regime's perspective, wilderness is a domain of uncaptured and therefore "monstrous" fecundity that has not yet been *purified* (to use Bruno Latour's schema in *We Have Never Been Modern*) into socially useful separate spheres of pastoral gardens and exploited colonies. As Aunt Elizabeth's comments reveal, Gilead's brutal "solution" to the problem of uncontrollable fertility, whether it be nature's or women's/trans-persons' is to eliminate such threats (via ecocide or execution). Hence, the hypocrisy of Gilead's "right to life" stance on abortion becomes clear: only life that complies with its dictates is worthy of protection.

Ultimately, *The Handmaid's Tale* and *The Testaments* illustrate that Gilead's attempt to control women's reproduction, to eliminate LGBTQ sexualities, to racially segregate and exploit persons of color and to dominate nature are all spectacular failures. Its attempts to purify sexuality and dictate reproduction result in a mass exodus of women from its domain and to covert corruption whereby highly placed patriarchs not subject to the same oversight commit heinous sex crimes. Its agricultural policies and exploitation of race-based slavery result in constant food shortages. And its production of toxic pollutants cannot be contained in colonial sacrifice zones but instead affect fertility rates inside Gilead. As Offred/Jane clarifies, only one in four births is viable because chemical and radioactive toxins "creep into your body, camp out in your fatty cells. Who knows, your very flesh may be polluted, dirty as an oily beach, sure death to shore birds and unborn babies" (112). Gilead's patriarchal refusal to acknowledge that *men* (as well as women) may be rendered sterile by such pollutants, and that this sterility results from the regime's militarism, reliance on extractive fossil fuels (signified by bodies that are as "dirty as an oily beach") and other unsustainable policies undermines its attempts to control reproduction and to secure a viable future. Although overpopulation is not the challenge that Gilead mobilizes to address, its totalitarian oversight of women's reproduction and elimination of their legal rights (along with those of racial minorities and LGBTQ folks) as an assertion of control in the face of ecological chaos constitutes a frighteningly plausible blueprint for where Anthropocene politics might head in the future if we allow hard-won human rights to erode under fundamentalist autocracies. Although literary critics such as Harold Bloom have dismissed *The Handmaid's Tale* as a "gothic dystopia" (2) with little applicability to mainstream politics given (he claimed) that our "our media [was] move[ing] in an opposite direction from the Fascist Republic of Gilead" circa 2009 (vii), appreciation of the novel's prescience has only increased in post-Dobbs decision America. Given the popularity of the Hulu television series inspired by the novel and of extremist anti-abortion legislation in many American states, it is now common for pro-choice advocates to don Handmaids' red robes and white bonnets during protests. In a United States where school boards are attempting to ban Atwood's novel, where LGBTQ rights are under attack, and where states such as Texas are well on their way to realizing Gilead's theocratic panopticon,[4] the resonance of Atwood's *Handmaid* books will only continue to grow.

Pregnancy as a 'Wilderness of Being': *Future Home of the Living God*'s De-colonization of Reproductive Rights and Deconstruction of Social Darwinism

Although Atwood's *The Handmaid's Tale* broke new ground in the 1980s by exploring parallels between women's oppression and ecological crisis, from a contemporary perspective the critique of reproductive rights undertaken in this text (and in its sequel *The Testaments*) can be seen as incomplete on at least three levels. First, scholars have critiqued a problematic "racial erasure" or a color-blindness in these novels, arguing they fail to acknowledge the *particular* challenges faced by women of color *vis-à-vis* reproductive rights by focusing exclusively upon the struggles of white women in Gilead, and (furthermore) that these texts suggest a problematic equivalency between this demographic's oppression under white supremacy and that of African American women under slavery. Second, the environmental concerns raised in *The Handmaid's Tale* are largely anthropocentric—focusing on the deleterious effects of environmental toxins on human reproductive capacities and on food production but rarely acknowledging the effects of such pollution on more-than-human agents. Though the animal-themed costumes of prostitutes at Jezebel's hints at links between women's and animal's exploitation in Gilead, this theme is not pursued in detail as it is in Atwood's MaddAddam books, nor is it linked with speciesism and the problem of mass extinction as it is in the later trilogy. Likewise, though *The Testaments* occasionally gestures at concerns about biodiversity, the most memorable anxieties about extractive industries (such as clear-cutting and mining) depicted in the novel are those that draw attention to the plight of women as "resources" for exploitation, not the suffering of animals displaced or eradicated by such industries. Third, because of the underrepresentation of women of color and more-than-human agents as victims of social and environmental exploitation, there is a lost opportunity in the *Handmaid* books to trace important intersections between women, people of color and animals as ideologically "othered" subjects who are all imbricated within colonialist and white supremacist reproductive politics. By focusing on the relatively narrow historical manifestations of gender oppression in totalitarian/fundamentalist regimes (from the time of Puritan colonists in the US onwards) and bracketing how Darwinian interpretations of longer, *evolutionary* history came to play a crucial role in buttressing white colonialism from the mid-nineteenth century onwards, the *Handmaid* texts don't provide a thorough critique of the pseudo-scientific biological essentialism that underwrites regimes such as Gilead. In the years following Darwin's *On the Origin of Species* (1859) and *The Descent of Man* (1871), proponents of social Darwinism conflated biological and social evolution in a telos that was used to justify white supremacy, imperialist expansion, sexist gender roles and oppressive reproductive politics in the name of eugenics. Although there's a general consensus among scholars who study the history of social Darwinism's influence in the West that this ideology lost its mainstream legitimacy following the "modern synthesis" of evolutionary theory and after World War II exposed the horrific outcome of race-based essentialism in the Holocaust and other Nazi atrocities, there's evidence of a contemporary resurgence of elements in this ideology in far-right politics. Indeed, as I shall examine, Alt-right "great replacement" theory perpetuates the racial essentialism that was a hallmark of social Darwinism, and (moreover) has forged alliances with white nationalist elements in the anti-abortion movement. Louise Erdrich's *Future Home of the Living God* brilliantly deconstructs the afterlives of social Darwinism and reproductive totalitarianism in the Anthropocene era, exploring how these forces mobilize in a near-future biopolitical regime that seizes power due to instabilities caused by climate change and the sudden onset of reproductive abnormalities that affect both human and animal populations. By focusing on how autocratic reproductive policies disproportionately impact women of color and by exposing the ways that evolutionary theories of socio-biological "progress" hinge upon racial and gender oppression, Erdrich's novel offers hope that the world might one day become "home" to ethnically diverse, sexually emancipatory societies united by trans-species kinship rather than divided by fear and hatred.

Given that Erdrich shares many of Atwood's concerns about women's reproductive rights, ongoing environmental problems and the potential rise of neo-totalitarian governments as a response to socio-ecological crisis, it is not surprising that reviews of *Future Home of the Living God* often compare Erdrich's dystopian novel with *The Handmaid's Tale*. Indeed, in a 2017 interview for *Elle* magazine with questions devised by Atwood, Erdrich discussed how she initially started *Future Home* as a response to the 2000 US Election, which she saw as a "disaster for reproductive rights" insofar as President George W. Bush reinstated the global gag rule which prohibited contraception funding for countries that offered abortion as a reproductive service. In this interview, Erdrich criticizes such measures at a time when the world "face[s] overpopulation," underscoring the potential ecological as well as social costs of such policies, concerns echoed in her sense that Bush's election meant losing a vital window for taking action on climate change. After setting aside *Future Home* for a number of years to pursue other works, she resumed it following the 2016 election, turning again to fiction as a means of processing President Trump's ascendancy and suggesting that the novel might be an opportunity to explore "the biological equivalent of our present political mess." Indeed, as many critics have noted, *Future Home* is an unapologetically *messy* novel, one that—despite the genre DNA it shares with other dystopian works—leaves much regarding the *causes* of the reproductive crisis that it represents as well as the *effects* of this crisis (include the logic behind governmental containment strategies) open to the reader's interpretation.

Cedar Hawk Songmaker, a young woman of Ojibwe and white descent who is the novel's primary protagonist, opens the narrative with the first of many diary entries addressed to her as yet unborn child. She relates that at the end of July in an unspecified year in the near future evolution suddenly stops behaving in the way we've become accustomed to and creates a sense that "our world is running backward. Or forward. Or maybe sideways, in a way as yet ungrasped" (3). Long-extinct species such as Archeopteryx and Saber-toothed cats are being re-born and human babies are (likewise) exhibiting developmental abnormalities in things such as neocortex formation. Whether these evolutionary changes can be accurately defined as a form of "devolution" is debatable because, as a paleontologist speaking about deep time relates on a television program watched by Cedar and her adoptive parents Sera and Glen Songmaker, neat linear or teleological concepts of evolution don't capture elements of chaos/ random mutation in such processes. As this scientist puts it,

> We do not have a true fossil record of human evolution … or any other species' evolution for that matter. What we have are bits and pieces that have survived and surfaced over millions of years … if evolution is going backward … then we would not see the orderly backward progression of human types that evolutionary charts are so fond of presenting. Life might skip forward, sideways, in unforeseen directions. We wouldn't see the narrative we think we know. Why? Because there was never a story moving forward and there wouldn't be one moving backward … We might actually see chaos.
>
> (54–5)

Here is the first indication that Erdrich will be challenging the stories we tell about evolution in the public sphere: namely, that it is a process of linear and hierarchal "progress" from lower to more advanced organisms. As we shall see, such misperceptions transferred to socio-political discourse have historically fueled regressive social Darwinist and colonialist agendas.

Roughly contemporaneous with the appearance of novel evolutionary "chaos" in Cedar's society is news that the US government is about to declare a state of emergency and is debating whether to quarantine members of the public due to a pathogen that may have been released from melting permafrost associated with climate change (8). Significantly, when Cedar's adoptive parents deliver this news to her via a phone conversation it is fragmentary and rife with uncertainty—exhibiting a communicative chaos that parallels the gaps and randomness in evolutionary processes:

Glen: Cedar, things are taking a more ominous turn, though I don't think people realize it yet … Listen to the news. There's a lot about …
Cedar: What?
Glen: The president is talking about declaring a state of emergency and there's a debate in Congress about confining certain …
Cedar: Dad, you're always—

[…]

Sera: Well, we don't know. This could be a new kind of virus. Maybe bacteria. From the permafrost. Use hand sanitizer, okay?

(8)

As readers, we do not know whether there's a causal link between the pandemic and the evolutionary crisis, though their temporal coincidence suggests this may be a possibility. In turn, both of these disruptions may (or may not) be related to the larger ecological instabilities created by a climate-altered world, which, by the time Cedar begins her journal, has already witnessed the end of snow and truly cold winters. In the face of uncertainty and spiraling environmental chaos, restrictions on women's reproductive rights become more and more draconian. One of the first things Cedar tells us is that "whatever is actually occurring, there is constant breaking news about how it will be *handled*" (3; my emphasis). Such responses range from technocratic attempts to "turn this thing around with gene manipulation" (54) to the "gravid female detention" policy eventually authorized by Congress wherein all pregnant women are forcibly "sequestered in hospitals in order to give birth under controlled circumstances" (72) to the even more oppressive detention and enforced insemination program mandated by Church of the New Constitution, a theocracy that overthrows the US government. Although reviewers such as Anita Felicelli criticize the *Future Home's* "baroque" tendency to "simultaneously over- and under-explain" plot details, I am more convinced by scholars such as Silvia Martínez-Falquina who assert that the narrative's "lack of detailed information about the changing natural and political contexts is a strategic element" designed to create an aesthetic of "uncertainty … that requires the active participation in the construction of meaning on the part of the reader" and by her conclusion that this aesthetic choice is "endowed with strong ethical implications" (167). Cedar's (and the reader's) active negotiation of uncertainty is a performative alternative to knee-jerk assertions of reductive control on the part of opportunistic biotech corporations and reactionary political regimes in the novel.

Another related "ethical implication" of Erdrich's ethos of uncertainty explored by Martínez-Falquina is how the novel opens up space for reflecting upon the historical status of Native Americans as peoples whose cultural continuity was rendered uncertain under colonialist genocide, but who exhibit resilience and the ability to prosper even in the face of contemporary socio-environmental crisis. As aforementioned, Cedar has Ojibwe heritage and, upon learning she is pregnant, seeks to establish a relationship with her biological mother Mary Potts Senior ("Sweetie") by journeying to a reservation north of Minneapolis. Once there, Cedar encounters a tribal community sustained by businesses such as a casino, as well as a gas station owned by Mary's husband Eddy. Eddy is a Harvard PhD who attempts to reform his reservation's education system and who becomes a tribal leader in the years following the pandemic. Although he is philosophically inclined towards pessimism, he takes a long historical view on current social instability. When Cedar expresses anxiety that "the world is going to pieces," Eddy responds that "it's always going to pieces" and that, because "Indians have been adapting since before 1492" they will "keep adapting" (28). The novel's plot supports this prediction: as infrastructure engineered by white Americans is "sucking dry" (32) vast aquifers below the Minneapolis/St. Paul region in a desperate bid to compensate for climate change's depletion of fresh water, Ojibwe on Mary and Eddy's reservation stockpile non-perishable food and begin converting land into "one huge, intensively worked, highly productive farm" (226). With the onset of the pandemic and birth crisis, affluent urban whites abandon their vacation homes on land adjacent to the reservation, and (under Eddy's leadership) the tribal community takes back this land that was promised to the Ojibwe under their "original treaty" (214) but which was lost under "the Dawes Act of 1862 [which] removed land from communal ownership" (213). The reservation also secures its self-defense by seizing the National Guard arsenal at Camp Ripley, recruiting both Native and non-Native veterans and "decolon[izing] the uniforms of the militia" (227). In sum, the reservation experiences a renaissance in the midst of wider social disintegration in what remains of the United States. As such, the novel imagines a "fictional reversal of power in order to confront settler colonialism," thereby "challenging ... the idea of progress and ... the stereotypes around the foundational dichotomy of civilization and savagism" (Martínez-Falquina 171–2).

While I find such historically informed interpretations of the novel compelling and am (likewise) convinced by the case that Anna Kemball makes for *Future Home* as a text that critiques the subjugation of indigenous peoples to "biocolonial" medical practices and "interrelated forms of domination such as those which position female Indigenous bodies as 'new colonies'" (159), such readings rarely broach the question of why the novel often explores subaltern politics within the wider framework of deep biological and/or geological time. In an interview for the *Guernica* magazine, Erdrich stressed that she "wanted to write a book that was about evolution" in multiple senses: one that encompasses the broad "biological chaos" of species-level change, but that also addresses cultural and even personal change: "the evolution and understanding that begins to incur between mothers and daughters in the book." In becoming pregnant, Cedar becomes aware of her participation in ancient processes of life unfolding on the planet, and experiences herself as being "haunted" by "a shadow self," a long history of "genetic mishaps" or unexpressed genes that are "knitted in right along with [her] working DNA" ("Reproductive Nightmares" 107). What is widely labelled "reverse evolution" by media outlets in her society (via sensationalist graphics depicting "humanoid figures growing hunched as they walked into the mists of time, while in the background Beethoven's Fifth Symphony dissolves into a haunting series of hoots and squawks" (52)) may in fact simply be a product of these "ghost genes" suddenly expressing themselves. While in hiding from the "gravid female detention" policy, Cedar and Phil (the white father of her child) discuss why such an embodied history of chance mutation and "mistakes" is unacceptable to creationists who believe in intelligent design. But more deeply, the novel is critical not only of creationist teleology but of *any* attempt to interpret biological evolution through a teleological framework. Erdrich radically deconstructs the idea that evolution is goal-directed or linear, instead emphasizing the role that chance and chaos play in life's disorderly development. In particular, her novel is critical of social Darwinist interpretations of evolution as a hierarchical process which rewards genetically superior populations by virtue of their capacity to dominate other beings—a theory that "makes life on earth a scenario of bloody, ham-handed, ruthless, tooth-and-nail struggle" (107). For Cedar, "pregnancy is a wilderness of being"(67) that undercuts genetic mastery at every turn, including the US Government's and the Church of the Constitution's eugenic agendas. By showing connections between fallacious understandings of evolution, the afterlives of colonialism and how these things come together in oppressive reproductive policies, the novel not only underscores the inadequacy of technocratic "solutions" to Anthropocene challenges, but also exposes links between white supremacy and autocratic attempts to control women's reproduction.

At this juncture, it is important to specify key concepts associated with imperialist social Darwinism and its history of influence in the United States and Europe. Only by making this brief detour can we begin to appreciate Erdrich's deconstruction of this ideology's afterlives in *Future Home*. Although Gregory Claeys asserts that it is "a mistake … to presume that there was a single 'politics' of Social Darwinism or that all forms of Social Darwinism were illiberal" (228), he does acknowledge (following Mike Hawkins's *Social Darwinism and European and American Thought 1860–1945*) that there are four main assumptions that underwrite the

> Social Darwinist world-view: First, biological laws govern the whole of organic nature, including humans; second, the pressure of population growth on resources generates a struggle for existence among organisms; third, physical and mental traits confer an advantage on their possessors in this struggle … and last, the cumulative effects of selection and inheritance over time account for the emergence of new species and the elimination of others.
>
> (228)

In other words, despite differences in political applications, social Darwinists are united in "their use of quasi-biological or organicist explanations of social evolution, class divisions and poverty, and racial and national stages of development." Scholars such as Claeys and Thomas McCarthy challenge earlier accounts of social Darwinism as a sociological *appropriation* of Darwin's scientific thought in *On the Origin of Species* (1859) and *The Descent of Man* (1871), noting that "many of the ideas—on women and race, for instance—that Darwin's defenders often attribute to the *mis*use of his thought by ideologues unschooled in biology can already be found in Darwin" (McCarthy 75). These scholars examine the influence of pre-existing racial theories such as those associated with Manifest Destiny in the US and Victorian colonialism in Britain on both Darwin's evolutionary theory (particularly in *The Descent of Man*) and the work of social Darwinists such as British sociologist Herbert Spencer and his American disciple William Graham Sumner. Influenced by Malthus's theories on population, Spencer not only coined the term "the survival of the fittest" in an 1852 article (published before Darwin's *Origin*) but also famously used this concept to justify *laissez-faire* economics in which individuals are allowed to "freely" compete and only the fittest survive as the so-called "natural" order of things. It is important to stress that social Darwinists saw such struggles as an engine of "progress" whereby certain groups "improved" at the expense of others; for them, human social evolution had a clear "direction" from primitive to complex civilizations, just as biological evolution could be understood as a progress from simple to advanced lifeforms. Although Darwin rejected teleological understandings of biological evolution in *On the Origin of Species*, by the mid-1860s (influenced by thinkers such as Spencer, W.R. Greg and Francis Galton) he accepted the thesis that intelligence (rather than fecundity) was the primary criterion for "fitness" in human beings and "came increasingly to hope that the optimal outcome of human natural selection would be the triumph of 'the intellectual and moral' races over the 'lower and more degraded ones'" (quoted in Claeys 237). Thus, in the *The Descent of Man*, Darwin's views on race are often of a piece with Victorian-era imperialist expansion insofar as he now posits "race … as a determinate, independent factor in human evolution" (Claeys 238) and naturalizes the process by which so-called savage races are "replaced" by civilized ones by virtue of the latter's innate superiority. Thus, although elements of social Darwinism preceded Darwin's major publications, in the years following *On the Origin of Species* and *The Descent of Man* this supremacist philosophy was "lent [the appearance of] scientific legitimacy" (McCarthy 80).

It is hard to overstate how destructive social Darwinism was when it was linked with social praxis from the mid-nineteenth century to World War II. This ideology was repeatedly evoked to justify heinous injustice directed at people of color and women. For example, Rutledge M. Dennis's "Social Darwinism, Scientific Racism, and the Metaphysics of Race" discusses Sumner's justification of African American slavery on the grounds that because this institution "permitted superior groups the leisure to construct and develop more refined cultures, it actually advanced the cause of humanity" (244). Thomas McCarthy likewise discusses ways in which social Darwinism was employed to deny black voting rights after the Civil War and to support Jim Crow segregation in the 1890s on the grounds that "the inherent inferiority of blacks made them unfit for full citizenship" (77). In the sphere of gender politics, Darwin opposed the use of birth control[5] and, in *The Descent of Man,* argued that evolution made men intellectually and artistically superior to women, views which influenced Spencer's assertion that for the human species to progress women had to devote their lives to reproduction.[6] Although Kimberly A. Hamlin's *From Eve to Evolution: Darwin, Science and Women's Rights in Gilded Age America* documents the fact that feminist intellectuals (both women and men) used evolutionary theory to contest such sexism and to argue for gender equality, social Darwinist principles were often cited as grounds for denying women access to higher education and the right to vote. Furthermore, as aforementioned, social Darwinist thought was often employed as a convenient rationale for Euro-American colonialism in the nineteenth century. Although neither Spencer nor Sumner supported imperialism, the British mathematician Karl Pearson and the sociologist Benjamin Kidd both used social Darwinism as a justification for Euro-American expansion and viewed the colonization of non-caucasian countries/resources as "natural components of the Darwinist principles entailed in the struggle for existence, survival, and supremacy" (Dennis 245). Social Darwinist principles were also used to legitimate American imperial expansion during the Spanish-American war, as historian Thomas F. Gossett summarized:

> by the time of the Spanish-American war, the idea of race had deeply penetrated nearly every field ... [such that] Darwin himself had invested the westward movement of the American nation with all the force of a law of nature.
>
> (310–11)

For many social Darwinists, the genocide of indigenous peoples was an essential component of Euro-American colonialism as "it was widely assumed that 'inferior' races would fall victim to the onward march of progress" (Claeys 239). In the United States, events such as the massacre at Wounded Knee in 1890 were seen as evidence that Native Americans "were no longer a threat and seemed in fact destined to disappear, as social Darwinist views led many to expect" (McCarthy 77).

Erdrich's *Future Home* strongly refutes this predicted "destiny" of Native American peoples by representing the Objiwe's capacity to not only survive historical genocide but to constantly "adapt" and even thrive in challenging Anthropocene conditions. By deconstructing neo-social Darwinian accounts of historical and biological evolution in a near-future dystopia, Erdrich's text exposes this ideology's corrosive ongoing legacy in autocratic, patriarchal and white supremacist approaches to governance. It is not a coincidence that women's reproduction becomes a locus around which power dynamics mobilize as a response to ecological and ontological crisis in the novel. The text explores women's reproductive rights as a contested ground not only due to reactionary racist and sexist politics but also (critically) because reproduction at once links our species to others in evolutionary processes and, for social Darwinists and humanists alike, also *separates* us from other animals, constituting a basis for hierarchical, *species*-level domination as well as social oppression. In this sense, *Future Home* undertakes a more comprehensive critique of the ideology underwriting white supremacist patriarchy and anthropocentrism than Atwood's *Handmaid* books. Although Atwood's dystopian texts hint that something like social Darwinism underwrites the biological essentialism that the elite in Gilead employ to justify oppression, there is little attempt to systematically expose this essentialism's deeper ideological foundations in evolutionary science; rather, biblical authority is typically identified as the source of such essentialism. Unlike Atwood's novels, *Future Home* focuses on how women of color experience the intersection of gender- and race-based discrimination within the context of reproductive rights and how these forms of oppression can intersect with anthropocentric politics. Consequently, Erdrich's dystopia imagines a more durable basis for hope than Atwood's books that multispecies justice may emerge in the future.

Importantly, it is often Cedar's contemplation of deep time that enables her to question teleological accounts of evolution, including social Darwinist assumptions about what constitutes "progress" in human societies or "fitness" in lifeforms more generally. For example, early in the novel Cedar's white step-parents argue in their air-conditioned home about the possible cultural impacts of the "reverse evolution" being reported in the news media. Sera bemoans what she feels will be the loss of markers for human achievement, exclaiming "Dear god … there goes poetry, there goes literary fiction, there goes science, there goes art" (55), but Glen points out that "[c]ave art was exquisite" and that "we have no idea of the capacity of our ancestors to think and feel," speculating that future non-sapiens hominins may be "intelligent." After hearing this debate, Cedar steps outside into an "unusually cool" 90-degree August afternoon and thinks about the deep history of life on earth. In doing so, she symbolically leaves the technological sanctuary that her privileged "trust-fund" (56) parents enjoy as members of the global north and experiences the ecological costs of their comfortable existence: namely, a climate-altered world of extreme weather suffered by millions of the world's poor. Gesturing towards these costs by situating Cedar at the crossroads of the global elite and subaltern peoples insofar as she also experiences her Native relatives' less-privileged lives on the reservation, Erdrich invites readers to re-assess the cultural assumptions at work in her step-parents' debate, particularly Sera's prediction that all cultural achievement is now foreclosed because our species will degenerate into "foraging apes … in only a few generations" (55).

In sharp contrast to her step-mother's linear celebration of human progress in the relatively short span of recorded history, Cedar emphasizes *continuity* between contemporary humans and those who were painting caves 14,000 years ago, a span of time that is enormous from Sera's historical perspective but is "not even an eye blink" within geological timeframes. Imaginatively excavating the history of our species, from proto-humans' evolution about 4.4 million years ago to the emergence of "*Homo sapiens* … something like 300,000 years ago," Cedar is struck by the fact that the typical human being does "not understand how much time has passed on this planet" and has "no concept of our limited place in the enormousness of that time" (57). Given that species with limited cognitive capacity such as dinosaurs "lasted so much longer than we have," Cedar comes to question Sera's Eurocentric definition of human intelligence as an index of cultural "progress," something that is also central to social Darwinism's teleological account of human development. Cedar asks a radical question: could our species' "level of intelligence … be a maladaptation, a wrong turn, an aberration" rather than a marker of superiority? Such questions take on particular importance within the context of western colonialist assumptions of cultural superiority *vis-à-vis* indigenous cultures. Cedar's questioning of "intelligence" as a marker of "fitness" undercuts Darwinian hubris that would justify the genocide of Native Americans (and other non-western peoples) on the grounds that modern industrial society is inherently "superior" or more "fit" in the struggle for survival than primal cultures. From an environmental perspective, western colonialism can be seen as *regressive* in its domination/exploitation of natural resources, a "maladaptation" that, in the era of global neoliberal capitalism, has become a hegemonic force that has resulted in climate destabilization and cascading ecological crisis. As aforementioned, Cedar's thoughts on evolution within the context of deep time repeatedly acknowledge an irreducible indeterminacy in life's unfolding on the planet, such that she finds herself hoping that her unborn child might be similar to "Homo neaderthalensis" (55), not as a genetic throwback but as the emergence of a new kind of hominin hybrid (one that blurs the line between *Homo sapiens* and Neanderthals as explored in Bradley's *Ghost Species*). Significantly, when Cedar sees a "lizard-like bird" resembling "Archaeopteryx or something like it" in her backyard while she attempts to evade capture by government operatives, she doesn't see this creature as embodying a genetic regression but rather heralds it as a novel being, thinking, "I have that sense of time folding in on itself, the same tranced awareness I experienced in the ultrasound room. I realize this: I am not at the end of things, but the beginning" (92).

Unfortunately, Cedar's open-mindedness is not the dominant reaction to dramatic evolutionary changes going on around her. The US government passes a series of ever-more-invasive biopolitical interventions designed to manage what it sees as widespread reproductive degeneracy, beginning with declaring a state of emergency, then extending the "Patriot Act" to "seize entire library and medical databases in order to protect national security" (72), which allows agents to locate pregnant women for "gravid female detention." The novel suggests that women of color (particularly those in mixed-race relationships) are targeted for this detention, suggesting that the government has a racial- as well as species-oriented agenda. For example, as a woman with Ojibwe heritage who is having a child with a white man, Cedar is urged by her physician to go into hiding and not reveal that she's pregnant. Likewise, she later witnesses a horrific "gravid female detention" involving two police officers that wrestle a pregnant African American woman into their car while her traumatized white husband and daughter are powerless to stop this. Once Cedar is captured herself and put in a detention center, she discovers that her roommate Tia Jackson (aka "Spider Nun") is an Asian woman who (like herself) became pregnant with a white partner. As Cedar and Tia conspire to escape by weaving yarn from re-purposed blankets into a ladder—a practice reminiscent of her grandmother's Ojibwe finger-braiding—Cedar feels a sense of deep comradery with her roommate as a fellow woman of color attempting to escape from institutional oppression, noting that "when [Tia] tires, [she] start[s] working on the sash and from then on it is … [t]he two of us against them all" (141). Critics such as Sherryl Vint have analyzed the disproportionate "targeting and detention of racialized minorities" in *Future Home*, speculating why this might be the case given that the government's official rationale for imprisoning pregnant women is to preserve *Homo sapiens* as a *species*, a project that should (by this logic) be race-neutral. But as Vint astutely observes, the government's biopolitical agenda is far from neutral on this score; from their perspective

> The issue, at root, is not only that humanity survives into the future but also who is the exemplar of that humanity, thus engaging questions of the human as a racialized project revealing that anxieties about sterility are centrally anxieties about whiteness.
>
> (72)

In other words, when faced with a novel evolutionary crisis, US authorities shaped by neoliberalism and "colonial whiteness" replicate structural racial inequities in a eugenics program such that "concerns about the reproductive futurity of the species collapse into anxieties about the reproduction of current hegemonies" (85).

Although Vint's analysis focuses on how capitalist biotechnology in the novel replicates white supremacist and colonialist appropriation of women of color's reproductive capacities, she does not explicitly address how the US government's stockpiling of viable embryos and (later) the Church of the Constitution's program of forced insemination/ surrogacy of women are shaped by something resembling social Darwinism. While she addresses how "fitness" in Erdrich's dystopia becomes synonymous with "the capacity to accumulate genetic wealth" and quotes Eddy's prediction that "The wealthiest will get ahold of the technology to reproduce and those few *Homo sapiens* ... will own the rest of us, the monkeys" (95–6), she does not identify ways that this formula for who will survive replicates the historical emphasis on class- and race-based superiority as markers of "fitness" in Darwinian thought. As we have seen, non-whites were often dehumanized in such ideology as lower orders of being on the evolutionary scale, more akin to "monkeys" than culturally sophisticated Europeans and (thus) doomed to be "owned" by their superiors and/or eradicated by them in the struggle to survive. Thus, while Vint accurately notes that "the conviction that evolution is moving in some orderly, forward progression is completely exploded" (83) in *Future Home*, her analysis doesn't fully unpack what is at stake in this deconstruction: namely, exposing the pseudo-scientific evolutionary theory that (from the nineteenth century onwards) became a major justification for white supremacist colonialism and later "biopolitical" eugenics.

I would argue that the novel's engagement with social Darwinism provides a key to understanding the ironic *gap* between totalitarian governmental theories regarding the role that race plays in evolutionary changes afoot versus what's actually unfolding on a physiological level. Though Vint claims the novel suggests that "it [is] women of color whose pregnancies are most likely to exhibit genetic mutations giving rise to the next iteration of hominids" (80), I believe this is open to interpretation and that she may be conflating the authoritarian government's Darwinian (racist) assumptions with what is actually happening on a biological level. Ostensibly, the government justifies its bid to control women's reproduction on the basis of anxiety that *Homo sapiens* may be "replaced" by evolutionarily regressive hominins. Yet, the targeting of women of color—particularly those in mixed-race unions—for detention suggests another (ideologically related) anxiety may be at work: namely, that a homogeneously white population may be replaced by a racially hybrid one—insuring, from a white supremacist perspective, that human civilization is doomed to "regress" into a primitive state. What is at stake here (I would argue) is a conflation of species- and race-based hierarchies, one that bears more than a passing resemblance to Darwinian ideology and to contemporary "replacement" theory that has emerged as a toxic legacy of such views.

As Alexander D. Barder's *Global Race War* documents, French writer Renaud Camus's *The Great Replacement*, a text that has been influential for far-right anti-immigration politics in Europe and the USA and has inspired race-based terrorism in these countries, "fit[s] into a longer racial intellectual history in France that can be traced back to the 1890s" (226), including works by Édouard Drumont that take their cue from "scientific or biological racism of the nineteenth century" mixed with reactionary anti-Catholicism and anti-Semitism. While Barder also attempts to distinguish Camus's cultural (and Christian) essentialism from "the old scientific racism" of writers like Drumont (227), elements of replacement theory seem to replicate biological essentialism, particularly the concern that non-white reproduction in immigrant populations will outpace (and "replace") existing Euro-American populations, in effect winning the Darwinian struggle for reproductive fitness. Indeed, great replacement adherents like Camus share with social Darwinists "the premise that relations of domination and conflict necessarily come to define interactions between different races" (217) and a conviction that there are immutable differences which prevent non-white "others" from assimilating within western culture. Barder argues that the great replacement agenda is fundamentally an attempt to re-assert familiar, racialized "ontological boundaries" in a "genuinely postcolonial world," a project that "becomes about cementing in place the respective positions of different races/nations and rejecting the possibilities of a world on the basis of hybridity; especially when it is perceived that the wombs of non-Western women represent such an ontological threat" (230).

When viewed within the context of such reactionary ideology, the irrational governmental persecution of women of color involved in mixed-race relationships in *Future Home* takes on new clarity. From a perspective that insists on imposing clearly demarcated racial, cultural and species boundaries, women like Cedar represent a multi-dimensional threat insofar as they embody racial and cultural hybridity and evolutionary indeterminacy. The ugly assumption underwriting the detention of women like Cedar is that genetic and cultural homogeneity is the best prophylactic against a lapse into (in)human animality.

Holistically, *Future Home* brilliantly illustrates how misguided and dangerous such quests for purity and certainty are, both as a basis for understanding racial/cultural diversity and for comprehending the evolution of living beings. Though authoritarian governments depicted in the novel initially presume that the offspring of mixed-race couples are more likely (in Vint's words) to exhibit "genetic mutations" resulting in non-sapiens hominins, the text suggests that the reverse may (in fact) be true: i.e. that the children of such couples may be *less* apt to become something other-than-human. Significantly, the description of both Tia's and Cedar's babies do not suggest that they are (biologically) anything other than *Homo sapiens*. Though Tia's baby is stillborn, its face "looks like any baby, a crumpled little stone idol face, only a blue-gray color" (185). Likewise, when Cedar visits a gynecologist for an ultrasound, this physician assures her that her child doesn't exhibit any anomalies (is not a "throwback" (50)) and (on this basis) urges her to go into hiding, particularly when he learns that she has "special ethnicity" and is married to a white man. Ironically, though the government might assume otherwise, her fetus's measurements suggest a sapient identity that (given widespread mutations) has become extremely rare. The ambiguity of the clinic exchange could initially be read as suggesting that the "specialness" of Cedar's baby is that it is one of the new hominins; but this supposition is later undercut by the fact that there is nothing in her description of her newborn son that confirms this. Instead, what scant physical details are present—"velvety limbs … tensile and strong"; a newborn's "slaty blue" eyes "burning to live"; a tiny hand that holds Cedar's finger with "implacable strength" (264)—all suggest something at once human, more-than-human and divine. She describes her child's lanugo-covered limbs as "glow[ing] like copper" and sees "the soul of the world" when looking in his eyes. Clearly, the birth is a momentous spiritual event from Cedar's perspective, one that combines elements of Catholic as well as Native American religious iconography. The fact that Cedar's son is born on Christmas day and that her own given name is "Mary" resonate with Christian theology; but Cedar's "oceanic" experiences of labor (feeling she is surrounded by the "spirits" of earth, water, sun, moon and stars and hearing "thousands of spirits and human beings singing") suggest a pantheism more in keeping with her Ojibwe ancestry.

Though a *homo-sapiens* child, Cedar's offspring is an awe-inspiring hybrid on the deepest levels; a living chorus that gathers diverse genetic, racial and spiritual ongoingness. In a final, heart-breaking irony, Church of the Constitution officials confiscate Cedar's son and the novel ends with her facing either death (due to heart damage) or, like other women of child-bearing age, repeated forced insemination. It is possible that, given Cedar's child proves to be one of the so-called "originals," the Church's autocratic regime will try to use her (like other imprisoned women) as a vessel for the birth of a "future" savior who they hope will reverse profound evolutionary changes underway. Indeed, billboards such as *Future Home of the Living God* that Cedar sees when driving to the reservation seem to reflect such millenarian yearnings in the midst of apocalypse. Yet the text suggests that this regime (like other autocracies) won't recognize that the "living god" has always already *arrived* and that its proper "home" has always been the "fallow and weedy" earth—a limitless "horizon" that patriarchal and racist purists mistake for "one enormous, empty field" (13) awaiting human cultivation, just as they reduce women to empty vessels awaiting insemination.

Moreover, Erdrich's novel goes beyond merely deconstructing social Darwinist evolutionary teleology to suggest alternative ways of understanding human history within the broader context of life's deep history on planet earth. Bridgitte Barclay's "'My heart slowly cracks': Making Kin and Living through Extinction in Erdrich's *Future Home of the Living God*" sees the text as modeling humanity's evolutionary entanglement with other species along the lines explored by Donna Haraway in *Staying with the Trouble.* As evidence of this, Barclay reads Cedar's initial hope that her unborn child be Homo neanderthalensis rather than Homo erectus as reflecting a deep sense of kinship with non-sapient beings. Similarly, she reads the novel's focus on Native American culture's resilience in the face of white colonialist genocide as signifying hope for cultural continuity, for enduring societies that can adapt to ecological changes rather than imposing rigid teleologies onto human and more-than-human actants. She interprets Cedar and Tia's unweaving and re-weaving of fabric in order to create a lifeline (rope) that will enable their escape from patriarchal/eugenicist confinement as a metaphor for evolution's ongoing patterns of life, extinction and new life such that the novel as a whole can be seen as "a praise of material entanglements, assemblages, and ongoingness" (81) despite the fact that it also suggests that *Homo sapiens'* extinction may be immanent. Yet while such hopeful elements exist in Erdrich's text, Barclay's emphasis on these things tends to gloss over less optimistic elements therein. For example, if the novel is intended to be a "love letter to another species" as Barclay claims, why does Erdrich choose to make Cedar's baby an "original" (i.e. not a novel, non-sapient child), and why is Cedar's final meditation on her last memory of snow similarly nostalgic (backward-looking) regarding the old world that is being lost, rather than welcoming of the new world being born? Moreover, why does Barclay ignore the disturbing agency of microbial entanglements in the novel—i.e. the possibility that changes to animal DNA/reproduction may have been triggered by a virus or bacteria released in permafrost melted by climate change, a phenomena that Glen (and later Cedar) attributes to "Mother Nature [having] a clear sense of justice. You fuck me up, I fuck you up" (54)?

For all its power as a critique of patriarchal regimes that deny women's reproductive rights and of the faulty evolutionary ideology that underwrites supremacist praxis, *Future Home* is incomplete in its analysis of the economic hegemony that underwrites fossil-fuel capitalism. Although Eddy realizes that economically privileged groups in US society may be able to reduce their vulnerability to evolutionary changes afoot via biotechnology and (in contrast) sets his own reservation on a course to withdraw from capitalist hegemony to embrace independent living as a key to survival, his community nonetheless relies on fossil fuels as a primary power source and income generator:

> The gas pumps at the Superpumper are still being resupplied. The oil company could care less who's in charge. The casino was, of course, rich in cash. They still have reserves they have not touched, as does the tribal bank. (227)

Stated immediately after Eddy's avowed goal of making his people "self sufficient, like the old days," this information appears to undercut his sense that total secession from capitalist petromodernity is possible. Given there's no evidence in the novel that the reservation tries to build a new infrastructure based on renewable energy and ample evidence that neoliberal capitalism is in crisis, one wonders how long Eddy's community can rely on oil being re-supplied ad infinitum and why they aren't exploring green alternatives, particularly if their aim is long-term "self-sufficiency." Indeed, Eddy isn't the only character in *Future Home* whose ostensible rejection of ecologically destructive capitalism is compromised by his underlying reliance on this system continuing to function. Cedar's step-father Glen is an environmental lawyer who lauds nature's "just" retribution in the evolutionary crisis, and her step-mother Sera (another lawyer) is an anti-vax mom and advocate of "natural" home birth. Yet Sera sees an opportunity to profit from the crisis by "investing in one of those genetic companies" which she anticipates will "try to turn this thing around with gene manipulation" (54), and both her parents' avowed liberalism is cushioned by an economic safety net. In Cedar's words, her adoptive parents

> rely on substantial trust funds, which they shifted to bonds way back before the technology stock bubble burst the second time, then shifted out again, into real estate, then flipped their houses just before the last housing crash. Which is to say, they are shrewd as only market-based-society suspicious trust-fund liberals can be.
>
> (56)

It could be argued that because "even [the] combined ingenuity" of both Cedar's Native and white families "is not going to get [her] out of" (259) the Stillwater Birthing Center where she is confined at the end of the novel, Erdrich's text ultimately emphasizes the *illusion* of safety for women under neoliberal regimes, even those from relatively privileged backgrounds. Yet while the novel amply illustrates the dangers of denying women reproductive freedom in the face of Anthropocene crises ranging from climate change to pandemics, it leaves many questions about capitalism's economic entanglements with such phenomena unanswered. It is to such entanglements that we must now turn in the next chapter.

Notes

1 For more on current and projected population growth, see the United Nations.org website, "Global Issues: Population."
2 See Clark's *Ecocriticism on the Edge*, Haraway's *Staying with the Trouble*, and Bergthaller and Horn's *The Anthropocene: Key Issues for the Humanities.*
3 For Atwood's response to such prohibitions, see "Go Ahead and Ban My Book" in the *Atlantic* magazine, February 2023.
4 In 2021 Texas banned all abortions after six weeks (with no exceptions for rape or incest) and also extended programs like Pastor Mark Crutcher's "Spies for Life" mobilization of citizenry to enforce the new law, establishing a "bounty hunter" system whereby citizens in Texas (and other states) are given financial incentives ($10,000 plus legal fees) in order to sue anyone they suspect of facilitating an abortion (including doctors, insurance companies or Uber drivers taking women to clinics). For more on this, see Feuer.
5 See Darwin's 1877 "Letter to Charles Bradlaugh," wherein he discusses his opposition to birth control.
6 For more on Darwin's views regarding gender and how Antoinette Brown Blackwell took issue with gender theories in his work, see Rebekkah Rubin's "The Woman Who Challenged Darwin's Sexism."

References

As Women Worldwide Still Struggle to Achieve Basic Rights, Third Committee Emphasizes Importance of Access to Citizenship, Education, Work, Justice. United Nations, October 3, 2023, https://press.un.org/en/2023/gashc4375.doc.htm.

Assessing the State of Public Opinion Toward Women, Gender, Equality—and Abortion. PerryUndem, January 31, 2023. https://perryundem.com/wp-content/uploads/2023/01/PerryUndem-Landscape-of-Views-toward-Women-Gender-and-Abortion.pdf.

Attenborough, David. *A Life on Our Planet: My Witness Statement and a Vision for the Future*. Grand Central, 2020.

Atwood, Margaret. Dire Cartographies: The Roads to Ustopia. In *In Other Worlds: SF and the Imagination*, Anchor, 2012, pp. 66–96.

Atwood, Margaret. Go Ahead and Ban My Book. *The Atlantic*, February 12, 2023. www.theatlantic.com/ideas/archive/2023/02/margaret-atwood-handmaids-tale-virginia-book-ban-library-removal/673013/.

Atwood, Margaret. *The Handmaid's Tale*. Anchor Books, 1998.

Atwood, Margaret. *The Testaments*. Nan A. Talese, 2019.

Atwood, Margaret and Louise Erdrich. Inside the Dystopian Visions of Margaret Atwood and Louise Erdrich. *Elle*, November 14, 2017. www.elle.com/culture/books/a13530871/future-home-of-the-living-god-louise-erdrich-interview/.

Barclay, Bridgitte. 'My heart slowly cracks': Making Kin and Living through Extinction in Erdrich's *Future Home of the Living God*. In *Fiction and the Sixth Mass Extinction: Narrative in an Era of Loss*, edited by Jonathan Elmore. Lexington, 2020, pp. 67–86.

Barder, Alexander D. *Global Race War*. Oxford University Press, 2021.

Bergthaller, Hannes and Eva Horn. *The Anthropocene: Key Issues for the Humanities*. Routledge, 2020.

Bloom, Harold, editor. Editor's Note and Introduction. In *Margaret Atwood: Modern Critical Views* (New Edition). Chelsea House, 2009.

Claeys, Gregory. The "Survival of the Fittest" and the Origins of Social Darwinism. *Journal of the History of Ideas*, Vol. 61, No. 2, April 2000, pp. 223–40.

Clark, Timothy. *Ecocriticism on the Edge: The Anthropocene as a Threshold Concept*. Bloomsbury, 2015.

Darwin, Charles. Letter to Charles Bradlaugh, June 6, 1877, www.darwinproject.ac.uk/darwinletters/calendar/entry-10988.html.

Delacroix, Céline. Ecology and Empowerment. Great Transition Initiative, August 2022, https://greattransition.org/gti-forum/population-delacroix.

Dennis, Rutledge M.Social Darwinism, Scientific Racism, and the Metaphysics of Race. *The Journal of Negro Education*, Vol. 64, No. 3, Summer 1995, pp. 243–52.

Devall, Bill and George Sessions. *Deep Ecology: Living as if Nature Mattered*. Gibbs Smith, 1985

Erdrich, Louise. *Future Home of the Living God*. Harper Perennial, 2018.

Erdrich, Louise. Louise Erdrich. Reproductive Nightmares, Real and Imagined. Interviewed by Joanna R. Demkiewicz. *Guernica*, November 20, 2017, www.guernicamag.com/louise-erdrich-reproductive-nightmares-real-imagined/.

Felicelli, Anita. Louise Erdrich's Dystopian Dreams in "Future Home of the Living God". *Los Angeles Review of Books*, December 12, 2017, https://lareviewofbooks.org/article/louise-erdrichs-dystopian-dreams-in-future-home-of-the-living-god.

Feuer, Alan. Texas Abortion Law Creates a Kind of Bounty Hunter. Here's How it Works. *New York Times*, September 10, 2021, www.nytimes.com/2021/09/10/us/politics/texas-abortion-law-facts.html.

Foreman, Dave. *Man Swarm: How Overpopulation is Killing the Wild World*. Live True Books, 2015.

Gaard, Greta and Lori Gruen. Ecofeminism: Toward Global Justice and Planetary Health. In *Environmental Philosophy: From Animal Rights to Radical Ecology*, 4th edn., edited by Michael E. Zimmerman et al., Pearson Prentice Hall, 2005, pp. 155–77.

Gilman, Charlotte Perkins. The Yellow Wallpaper. In *The Norton Anthology of Short Fiction*, 7th edn. (shorter version), edited by Richard Bausch and R.V. Cassill, Norton, pp. 302–13.

Global Issues: Population. United Nations, www.un.org/en/global-issues/population.

Gossett, Thomas F. *Race: The History of an Idea in America* (New Edition) [1963]. Oxford University Press, 1997.

Hamlin, Kimberly A. *From Eve to Evolution: Darwin, Science and Women's Rights in Gilded Age America*. University of Chicago Press, 2014.

Haraway, Donna. *Staying with the Trouble: Making Kin in the Chthulucene*. Duke University Press, 2016.

Hawkins, Mike. *Social Darwinism and European and American Thought 1860–1945*. Cambridge University Press, 1997.

Heise, Ursula K. *Sense of Place, Sense of Planet: The Environmental Imagination of the Global*. Oxford University Press, 2008.

Hooker, Deborah. (Fl)orality, Gender and the Environmental Ethos of Atwood's The Handmaid's Tale. *Twentieth Century Literature*, Vol. 52, No. 3, Fall 2006, pp. 275–305.

Howells, Coral Ann. *Margaret Atwood*. 2nd edn., Palgrave Macmillan, 2005.

Human Population through Time. American Museum of Natural History, www.amnh.org/explore/videos/humans/human-population-timeline.

Iran: No End to Mounting Repression. Human Rights Watch, January 13, 2022. www.hrw.org/news/2022/01/13/iran-no-end-mounting-repression.

Issues: Abortion. Center for Reproductive Rights, https://reproductiverights.org/our-issues/abortion/.

Kauffman, Linda. Special Delivery: Twenty-first Century Epistolary in The Handmaid's Tale. In *Writing the Female Voice*, edited by Elizabeth Goldsmith, Northeastern University Press, 1989, pp. 221–44.

Kemball, Anna. Biocolonial Pregnancies: Louise Erdrich's Future Home of the Living God. *Medical Humanities*, Vol. 48, No. 2, January 17, 2022, https://mh.bmj.com/content/48/2/159.

Kuznetski, Julia. Disempowerment and Bodily Agency in Margaret Atwood's The Testaments and The Handmaid's Tale TV Series. *The European Legacy*, Vol. 26, Nos. 3–4, 2021, pp. 287–302.

Latour, Bruno. *We Have Never Been Modern*. Translated by Catherine Porter. Harvard University Press, 1993.

Malm, Andreas. In Wildness is the Liberation of the World: On Maroon Ecology and Partisan Nature. *Historical Materialism*, Vol. 26, No. 3, 2018, pp. 3–37.

Martínez-Falquina, Silvia. Louise Erdrich's Future Home of the Living God. *Atlantis*, Vol. 41, No. 2, December 2019, 161–78.

McCarthy, Thomas. *Race, Empire, and the Idea of Human Development*. Cambridge University Press, 2009.

Moffett, Paul. Erasing Race in The Handmaid's Tale. In *The Handmaid's Tale: Teaching Dystopia, Feminism and Resistance across Disciplines and Borders*, edited by Karen A. Ritzenhoff and Janis Goldie. Rowman and Littlefield, 2019, pp. 157–70.

Rubin, Rebekkah. The Woman Who Challenged Darwin's Sexism. *Smithsonian Magazine*, November 9, 2017. www.smithsonianmag.com/science-nature/woman-who-tried-take-down-darwin-180967146/.

Sethna, Christabelle. "Not an instruction manual": Environmental Degradation, Racial Erasure, and the Politics of Abortion in *The Handmaid's Tale*. *Women's Studies International Forum*, Vol. 80, 2020, www.sciencedirect.com/science/article/abs/pii/S027753951930 281X.

Vint, Sherryl. *Biopolitical Futures in Twenty-First-Century Speculative Fiction*. Cambridge University Press, 2021.

5 Viral Capitalism, Tangled Trees and the Future of Biodiversity

Can neoliberal economies adapt in time to mitigate multiple Anthropocene crises effectively—climate change, natural resource shortages, mass extinction and pandemics—for which fossil-fuel capitalism has been a major driver in the modern era? Such "mitigation" at the most basic level would entail changes aimed at averting global civilizational collapse and the legacy of an "uninhabitable earth" that David Wallace-Wells outlines in his 2019 book; or, more optimistically, the creation of economies wherein "sustainability" is re-defined to recognize humanity's irreducible entanglements with social, ecological and geological networks rather than the continuation of business as usual. Jason W. Moore poses similar questions in the conclusion to *Capitalism in the Web of Life*. Having developed convincing arguments for understanding capitalism as "*a way of organizing nature*" (2, emphasis in original) premised upon the availability of "Cheap Natures" (externalized ecological costs and unpaid labor) and the loss of such "cheaps" in the face of Anthropocene realities such as climate change, Moore asks whether we are "now living through a developmental crisis, one whose contradictions can be fixed through renewed capitalization, rationalization, and dispossession" or "an epochal crisis, one that will compel fundamentally new relations of wealth, power, and nature in the century ahead" (297). He inclines toward the latter, and casts doubt on neoliberal capitalism's capacity to solve Anthropocene problems via technological innovation, leaving open the question of whether "the exhaustion of the Cheap Nature model leads to something better, or something worse" in the future (294). But, in keeping with theorists such as Chakrabarty, Latour and others he acknowledges that global economies must reckon with the merger of historical and earth agency, asking, "[d]oes anyone today doubt that disease and climate make history every bit as much as any empire or class or market?" (296).

Within Anthropocene studies, "entanglement" has become a powerful model for theorizing the ontological and material enmeshment of human and more-than-human agents. Originally a concept in quantum physics, namely what Einstein described as the "spooky action at a distance" coupling of quantum particles, the term has come to mean something much broader in contemporary environmental theory. Entanglement's biological and/or ecological meanings point to an irreducible interpenetration of human and more-than-human actants such that "bounded individualism in its many flavors … has finally become unavailable to

DOI: 10.4324/9781003243946-6

think with, truly no longer thinkable" (5), as Donna Haraway puts it in *Staying with the Trouble*. In this work, entanglement is a trope that unifies what Haraway calls "SF" at different scales, ranging from quantum entanglement and string figures, to biological "sympoiesis" (processes whereby living things *make one another* rather than being autopoietic or self-constituting), to human/animal kinship networks, to speculative fictions/feminisms which weave new myths, stories of how to "stay with the trouble" in the face of contemporary ecological peril rather than resorting to familiar (but exhausted) narratives of technological triumphalism or paralyzing "game over" (3) apocalyptic despair. Likewise, Moore's analysis of capitalist *oikeios* assumes that entanglement is at the heart of "the relation through which humans act—and are acted upon by the whole of nature—in our environment-making" (4), yet it is this very enmeshment that capitalist ideology disavows through an ideological Nature/Society dualism that renders its reliance on cheap, plentiful natural resources and unpaid labor power invisible. Haraway agrees with theorists like Moore and Anna Tsing that the Anthropocene should be re-named as the "Capitalocene" or the "Plantationocene" to underscore the disproportionate role that capitalism and colonialism have played in bringing about our contemporary ecological crisis, yet her optimistic "Chthulucene" vision of the future in *Staying with the Trouble*'s "Camille stories" focuses primarily upon a revolution in kinship relationships, rather than on the issue of whether (or how) neoliberal capitalism can be transformed into a system that's compatible with multispecies environmental justice. Likewise, Anna Tsing's *The Mushroom at the End of the World* invests much hope for the future in the unpredictable emergence of "assemblages" or networks of "collaborative survival" involving human and more-than-human agents that manage to thrive even in capitalism's ruins. Tsing's primary examples of such resilient entanglement are ethnically diverse mushroom pickers who, having been unable to secure a living wage within neoliberal markets premised upon workers' economic precarity, gather spontaneously in places such as "Open Ticket" Oregon to harvest fungi that thrive in human-disturbed forests. Yet the avowed "freedom" that such labor entails is (as Tsing acknowledges) "irregular" and "effervescent" given that it takes place only in communities supposedly "outside rationalization" (76) where pickers claim they do not experience the alienation typical of workers exploited by neoliberalism and its destructive "salvage accumulation" (63) practices. As Marxist theory makes clear, however, workers can (in fact) be alienated without experiencing themselves as such; and the entanglement of Tsing's mushroom pickers within a cash economy that is vital to Matsutake supply chains suggests that their sense of "liberation" from neoliberalism is problematic. Like Haraway's "Camille stories," Tsing's celebration of resilient multispecies collaboration within specific "pericapitalist sites" is a laudable analysis of agency on a local/small-scale level, but doesn't give readers the tools to imagine what can be done to change neoliberal capitalism's ongoing social and ecological ruination at the macro (global) level.

A step towards developing a more thoroughgoing critique of the Capitalocene may be to engage more directly with the problem of the radical incommensurability between neoliberalism's temporal frames and those of the Earth's evolving, entangled ecosystems. For all its power as a means of mapping ecological enmeshment, new materialist notions of "entanglement" must come to recon with capitalism's distinctive *temporal* distortions. As Moore (quoting Marx) clarifies,

> capital incessantly drives towards the "annihilation of space by time" ... seek[ing] to create a world in which the speed of capital flows—its turnover time—constantly accelerates. The privileging of time over space in capital's project is not passive but active: every effort to accelerate turnover time implies a simultaneous restructuring of space. [This process is illuminated via] Lefebvre's powerful observation that capital not only occupies, but also produces, space.
>
> (10, emphasis in original)

Capitalism's privileging of time over space in an effort to maximize economic efficiency has had profound socio-ecological consequences, including a tendency of market economies to focus on short-term profit (benefits) rather than long-term sustainability (costs); but also, even more radically, a tendency for near-term "futures" speculation to "condition the present" as Jean-François Lyotard analyzes in *The Inhuman.* Lyotard's famous exploration of how late capitalist information technologies mobilize around the project of reducing future risk is worth citing in some detail:

> if one wants to control a process, the best way of doing so is to subordinate the present to what is (still) called the "future," since in these conditions the "future" will be completely predetermined and the present itself will cease opening onto an uncertain and contingent "afterwards" ... In this manner, the future conditions the present. *Exchange requires that what is future be as if it were present.* Guarantees, insurance policies, security are means of neutralizing the case as occasional, or as we say, to forestall eventualities ... Money here appears as what it really is, time stocked in view of forestalling what comes about ... nothing other than time placed in reserve, available.
>
> (65–6, my emphasis)

Lyotard's analysis implies that capitalism colonizes *time* as much as it does space, ordering temporality and seeking to minimize future risk (uncertainty) by treating future exchange *as if it were already present*, thereby emptying futurity of its radical alterity. In this way, timescales are contracted more radically than is the case with merely trying to maximize efficiency in the present; the future itself is treated as "standing-reserve" in a Heideggerian sense of something that must make itself "available" on-call to meet the demands of the present and, crucially, is knowable *only* via this enframing. Hence the market's myopic focus on quarterly data and near-term performance, but also its radical blindness to what theorists such as Rob Nixon have characterized as neoliberalism's "slow violence"—forms of environmental injustice that cannot be represented as spectacular singular events limited to the present time, but (rather) unfold over long temporal frames or linger in the environment for mind-boggling amounts of time. Thus, while neoliberal exchange takes place at an ever-accelerating pace that draws the future into an ever-contracting black hole of presentism, extractivist, fossil-fuel-powered capitalism is producing forms of pollution such as climate change that, as we have seen, stretch forward into future temporal expanses that dwarf our generation-bound ways of thinking. So the problem we face in critiquing late global capitalism is twofold: not only the question of how our economic systems might adapt to reflect a better understanding of their entanglement within broader earth-systems (the challenges Moore addresses in *Capitalism in the Web of Life*) but also the question of how we might *liberate time*—particularly futurity—from its reduction to destructive short-term thinking that underestimates the role of ecological (as opposed to merely economic) contingency.

On this score, literature that addresses our species' evolutionary entanglement with others across long expanses of historical and biological deep time restores this sense of futurity's radical alterity, or "wildness." This chapter will explore texts that represent two of humanity's most significant deep-time entanglements—our relationship with viruses and with trees—with the goal of illuminating neoliberal capitalism's shortcomings in mapping uncertainties that flow from these enmeshed relationships. Literary texts often illuminate such perceptual failure via "slow violence" that periodically erupts into full-blown crisis, from pandemics (in the case of viruses) to rapid die-offs, deforestation and destructive wildfires (in the case of trees). Inspired by David Quammen's *The Tangled Tree: A Radical New History of Life*, I will address how entanglement over deep time underscores the vital necessity that we develop an ethics of multispecies justice and economic systems flexible enough to adapt themselves to such an ethic.

My working thesis is that Anthropocene-era literary and theoretical texts expose neoliberal capitalism's viral temporality and material praxis, as well its destructive denial of both the benefits and risks arising from multispecies entanglement. As thinkers such as Haraway and Tsing make clear, forms of entanglement (such as symbiosis) should not be seen as a "synonym for 'mutually beneficial'" (Haraway 60). On a biological level, some forms of entanglement (such as those involved in infectious disease) are parasitic. On an ecological level, what Stacy Alaimo describes in *Bodily Natures* as the "transcorporeal" buildup of toxic chemicals within ecosystems and human/more-than-human bodies is also a mode of entanglement. Likewise, even forms of entanglement that currently sustain life such as the "wood wide web" (a 450-million-year old network of trees, fungi and bacteria) could, with sufficient disturbance from global warming, release "lots of heat-trapping carbon dioxide" from decomposing matter at an accelerated rate in the future, exacerbating climate change.[1]

All of which is to stress that "entanglement" is not something that excludes either viral or arboreal agency, but rather, encompasses both. In critiquing capitalism's "viral" agency, I will be focusing on how its temporal and spatial effects resonate with parasitic forms of entanglement, but this does not mean that symbiotic forms of viral entanglement do not exist. Indeed, as Quammen's *The Tangled Tree* reveals, recent discoveries in molecular biology show that the history of evolution on Earth can no longer be accurately represented merely by discrete branching lineages of vertical genealogical descent among members of the same species; rather, genomes are also shaped by horizontal gene transfer (HGT) or "infective heredity" whereby genes "can also pass laterally across species boundaries … even between different kingdoms of life" (xiii), as in the case of a retrovirus that made it possible for mammals to develop placentas and give live birth. Given that viruses are intrinsic to our makeup ("roughly 8 percent of the human genome consists of the remnants of retroviruses that have invaded our lineage" (349)) and can have both beneficial and destructive effects within our bodies and within ecosystems more broadly, it would be a mistake to stigmatize (or exclude) viral agency from what I am characterizing as "entanglement." To do so would be to replicate the kinds of wishful (and destructive) exclusions that have mobilized viral capitalism. Instead, as we shall see, what we need are new socio-economic models capable of acknowledging both the risks and splendors of a contingent evolution in which viral–tree–human entanglements are inextricable.

Viral Entanglements: Biocapitalism, Pandemics and Mass Extinction

There's a curious moment in James Bradley's climate change novel *Clade* (2017) when, in the midst of a global pandemic, a teenager named Li Lijuan under lockdown in rural Australia makes the following entry in her "journal of the plague year":

> NOVEMBER 8
> These are the things we've lost:
>
> Birds
> Bananas
> Tigers
> Frogs
> Bees
> Coffee
> Polar bears
> Coral
>
> These are the things we've saved:
>
> Seeds
> Elephants
> Dolphins
> Each other

This ledger of extinct and surviving species appears without prior context or explanation and, on a first reading, seems oddly out of place in Bradley's otherwise harrowingly prescient depiction of a novel respiratory virus that first appears in Guangxi China before rapidly spreading to the rest of the world. Yet a broader survey of contemporary cli-fi suggests a deeper pattern may underlie this seemingly strange entanglement of Anthropocene anxieties. Many critically acclaimed examples of Anthropocene-era literature reveal important links between neoliberal capitalism, climate change, increasingly frequent zoonotic diseases and rising extinction rates in non-human species, the latter often memorialized via lists of losses. For example, in Margaret Atwood's MaddAddam trilogy, a bioengineer named Crake (aka Glenn) plays an online game called "Extinctathon" wherein contestants try to guess the identity of animals or plants that have become extinct in the past 50 years due to his hyper-capitalistic society's environmental destruction. To give himself a competitive edge, Glenn consults a list of recently extinct species, a roster whose length (tellingly) "was a couple hundred pages of fine print" (81). As I will discuss later in this chapter, it is no accident that Extinctathon is Glenn's gateway into bioterrorism, not only influencing his symbolic allegiance with extinct bioforms when he re-names himself "Crake" but also his decision to create a deadly pandemic in order to add *Homo sapiens* to the list of no longer viable species.

A similarly grim accounting occurs in the opening of Peter Heller's eco-apocalyptic novel *The Dog Stars*. The protagonist Hig, survivor of a flu pandemic that "killed almost everybody" (9), begins his narrative with a "lamentation" not for his lost wife Melissa but with lists of extinct animals:

> If I ever woke up crying in the middle of a dream, and I'm not saying I did, it's because the trout are gone every one. Brookies, rainbows, browns, cutthroats, cutbows, every one. The tiger left, the elephant, the apes, the baboon, the cheetah. The titmouse, the frigate bird, the pelican (gray), the whale (gray), the collared dove. Sad but. Didn't cry until the trout swam upriver looking for maybe cooler water.
>
> (3)

As in *Clade*, no explicit causal link is drawn here between a pandemic that wipes out most of humanity and accelerating wildlife extinction rates, though Heller's narrative (like Bradley's and Atwood's) suggests that climate disruption (4) and associated environmental impacts (such as forest loss due to pine beetle infestation (5)) sets the stage for both disasters. Indeed, Hig's eulogy for extinct trout echoes the elegiac conclusion in Cormac McCarthy's *The Road* which mourns the irreplaceable loss not only of a habitable Earth for humans but also of a "world" in which "brook trout" still populated "deep glens where … all things were older than man and … hummed of mystery" (287).

Nor are the losses catalogued in Anthropocene plague fiction limited to animal species. Novels such as Paolo Bacigalupi's *The Windup Girl* also address the impact of human-influenced pandemics upon plant biodiversity. This text focuses on struggles to control food production in a future post-fossil-fuel Thailand. An American corporate spy named Anderson Lake attempts to locate seed banks hidden in the Thai kingdom in order to use this genetic material to create new genetically modified crops—an increasingly desperate gambit in a world whose food sources have been decimated by viruses that either develop spontaneously or which have been bioengineered to infect crops in competing markets. Novels such as *Clade*, those included in the MaddAddam trilogy and *The Windup Girl* expose how pandemics (such as Covid-19) arise from the same global economic forces that are accelerating extinction, producing climate change and widening social inequities. The slow violence of ecological problems accompanying the spread of sixteenth-century European colonialism and eighteenth-century fossil-fuel-powered capitalism manifests in phenomena that are often challenging to visualize, such as climate change, soil erosion, ocean acidification and rising extinction rates in more-than-human species; whereas deadly pandemics and weather-related disasters crystalize such violence in spectacular events effecting both humans and other species. In this way, the 2020 pandemic serves as a window onto capitalism's entanglement with viruses such as Covid-19 by way of factors such as deforestation and the destruction of wilderness habitat, the global trade in wild animals and climate change.

A number of prominent cultural critics have already written work linking Covid-19 to climate change. For example, Bruno Latour's "Is This a Dress Rehearsal?" develops the hypothesis that "the health crisis prepares, induces, and incites us to prepare for climate change" insofar as both crises lead us to "the sudden and painful realization" that in contemporary life the classical definition of society—humans among themselves—makes no sense.

> The state of society depends at every moment on the associations among many actors, most of whom do not have human forms. This is true of the microbes … but also of the internet, the law, the organization of hospitals, the logistics of state, as well as the climate… The pandemic is no more a "natural" phenomenon than the famines of the past or the current climate crisis.
>
> (S25–6)

Latour asserts that the Covid-19 pandemic, like climate change, can only be comprehended by modelling this crisis as the product of entangled (or "distributed") agency—the material and cultural effects of human and more-than-human material actors operating at different scales. Or, as Slavoj Žižek observes in *Pandemic! Covid-19 Shakes the World*,

> the coronavirus epidemic itself is clearly not just a biological phenomenon which affects humans: to understand its spread, one has to consider human cultural choices (such as food habits), economy and global trade, the thick network of international relations, ideological mechanisms of fear and panic. To properly grasp this link, a new approach is needed.
>
> (110)

Both Latour and Žižek share a sense that global problems such as Covid-19 and climate change demand a new materialist politics capable of overcoming modernist nature–cultural dualisms, and point to the inadequacy of governmental responses that construe these crises as "natural" or (in the case of Covid-19) as unavoidable "acts of God." As we shall see, a new politics premised upon human/more-than-human entanglement would illuminate how global trade in exotic animals as luxury food sources for the wealthy has led to increases in zoonotic disease transmission; but also, more generally, how commercial development of wilderness areas contributes to such transmission as wildlife is killed or displaced and viruses seek new human hosts as their animal "reservoir" hosts disappear. Global warming likewise increases the spread of diseases carried by mosquitoes and ticks, as warming temperatures allow these insects to expand the range of their habitats—a trend which (like viral pandemics) stands to affect humans and wildlife alike.

Literary texts such as those cited at the start of this section are critically important for raising public awareness about connections between climate change, disease vectors and reduced biodiversity. Furthermore, they also shed light upon ways that viral reproduction can be seen as global capitalism's uncanny double, the shadow form of an economic system predicated upon limitless growth, parasitic resource depletion and the spread of monocultures. Or, as Latour puts it in "Is This a Dress Rehearsal?," when it comes to global ecological change

> the pathogen whose terrible virulence has changed the living conditions of all the inhabitants of the planet is not the virus at all – it is humanity! But this does not apply to all humans, just those who make war on us without declaring war on us.
>
> (S27)

Similarly, in *Pandemic!* Žižek echoes Latour's sense that it is the global elite who benefit the most from capitalism who are waging such a "war," drawing an analogy to H.G. Wells's *War of the Worlds*:

> Perhaps an epidemic which threatens to decimate humanity should be treated as Wells's story turned around: the "Martian Invaders" ruthlessly exploiting and destroying life on earth are we, humanity, ourselves; and after all devices of highly developed primates to defend themselves from us have failed, we are now threatened "by the humblest things that God, in his wisdom, has put upon this earth," stupid viruses which just blindly reproduce themselves—and mutate.
>
> (13)

Despite the ostensible difference this passage tries to draw between Wells's technologically sophisticated Martian invaders who are destroying life on Earth (aka capitalist technocrats) and "stupid viruses" such as Covid-19 that threaten to "undo" this invading force, there is also a slippage at work here, insofar as these invaders parasitically "exploit and destroy life" in order to "blindly reproduce" capital, historically "mutating" at certain points given periodic technological revolutions, but fundamentally replicating the same exploitative economic system.

Such characterizations of neoliberal capitalism as a "viral" or "alien" force could be readily dismissed as merely metaphorical at best, or (more darkly) as offensively misanthropic reductions along the lines critiqued by defenders of humanism such as Murray Bookchin.[2] However, denying capitalism's viral entanglements outright instead of analyzing the material bases for such characterizations elides important connections between global capitalist praxis, fossil-fuel powered technology, climate change, the destruction of wildlife habitat, mass extinction and what Andreas Malm characterizes as "chronic" zoonotic pandemics in *Corona, Climate and Chronic Emergency*. Malm's analysis not only illuminates a "thick network of international relations" along the lines Žižek calls for as a heuristic for understanding the distributed agencies that contributed to the Covid-19 pandemic, but also shows why neoliberal capitalism's particular modes of appropriating both space and time create ideal conditions for future disease outbreaks.

The spatial dimensions Malm discusses include capitalism's role in global deforestation in areas such as Southeast Asia, Latin American and other tropical zones, destroying old growth forests and replacing them with soybean and palm oil plantations, or with grazing land for beef cattle. Such practices reduce biodiversity, creating isolated islands of wildlife in oceans of monoculture, and also encourage zoonotic disease transmission due to the aforementioned extermination (and/or stressing) of wildlife hosts. Likewise, the harvesting of wildlife for purposes of "exotic" luxury food consumption involves supply chain locales such as "wet markets" which are ideal breeding grounds for zoonotic pathogens insofar as caged animals are often stored in crowded, unsanitary conditions. Malm discusses bats as creatures who illustrate links between capitalism's conquest of space, habitat destruction, loss of biodiversity and zoonotic spillover events, including the sequence that may have triggered the 2020 Covid pandemic. In the wild, bats have remarkably strong immune systems which have developed "over millions of years" of co-evolution with pathogens such as coronaviruses; over time, they have become reservoir hosts for these diseases which "permanently inhabit … their bodies without killing them" (32). When bats are stressed due to loss of forest habitat, their immune systems become compromised and they shed more of these pathogens than they normally would (43); likewise, when they are caged and transported next to other wild animals like pangolins bound for "wet markets" the chance for zoonotic spillover greatly increases. When they, and other keystone wildlife species, are culled or rendered extinct in the name of "progress," the biodiversity that acted as a firewall or "dilution effect" (40) for naturally occurring pathogens is lost, replaced by opportunistic "weedy" species well adapted to living in close quarters with humans (think mice or rats) that often bring these pathogens with them.

Malm's analysis also makes clear how such dislocations brought about by capitalism's colonization of space are exacerbated by its technologically enabled appropriation and compression of time. Although zoonotic disease transmission is "an old story" (33) in human history (ancient as well as medieval trade routes spread plagues), fossil-fuel-powered transport in the post-industrial-revolution era exponentially sped up the *rate* at which pandemics could spread around the globe. Likewise, some ecological studies (such as one titled "The Coevolution Effect as a Driver of Spillover Effect" cited by Malm in *Corona, Climate and Chronic*

Emergency) have suggested that islands of ecological biodiversity in fragmented landscapes may accelerate pathogen evolution such that "the engine of deforestation acts to rev-up spin-off engines of parasitic experimentation" (41–2). In sum, one of the most remarkable "hypotheses" that comes out of Malm's detailed study of how neoliberal networks of exchange facilitate disease transmission is that capitalist praxis as a whole strongly resembles the parasitic replication it enables. In a section titled "Towards a theory of parasitic capital," Malm observes that the history of market growth and the conquest of wilderness suggests that

> Capital can relate to the wild *only by attaching itself to it*, so as to make it give up commodities that bear exchange-value; and in the very moment of its success, that nature is wild no more: it is clear-cut, captured, caged and carried off to market. It is eaten from within, but the damage is inadvertent. Capital doesn't *mean to* destroy the intricate cellular structures of wild nature; it doesn't have an intention formed in the mind and then engage[d] in efforts to realise it—there is just no other way for it to replicate. The fastening and sucking are in its DNA (or is it RNA?); the moment they cease, the reproduction of capital is over. Unlike other parasites, this one cannot stay content with vegetating in the furs or veins of other species for millions of years of co-evolutionary equilibrium. It can subsist solely by expanding and, in this sense, it exhibits a sort of permanent pandemicity.
>
> (77–8)

There is much to unpack in this passage, but most apropos for our exploration of viral capital is this system's zombie-like self-replication and mindless (non-intentional) destruction of living systems. Although Malm acknowledges that such a theory of "capital as meta-virus and patron of parasites" must be backed up with more in-depth research, he offers the less abstract theorem that capitalism's "time-space appropriation plus time-space compression equals high risk of zoonotic pandemics" (78). Drawing upon Alf Hornborg's theory regarding capitalism's "time-space appropriation" and David Harvey's ideas regarding its "time-space compression," Malm clarifies that the former entails market-driven attempts to locate areas that offer what Moore would term both "cheap nature" and "cheap labor," whereas the latter involves using technology to speed up market efficiency: "the faster an investment can be recouped, the quicker commodities are sold and return their yield, the greater the profits. Capital grows by higher velocity as well as higher volumes" (80). Hence, Malm's sly aside regarding parasitic capital's possible replication via RNA rather than DNA, an analogue to coronavirus's capacity to evolve more quickly than other pathogens via RNA mutation.

It must be emphasized that in analyzing capitalism's viral entanglements one should *not* conclude that it is a "natural" or inevitable system, nor that human beings in general are "hard-wired" to behave like viruses. In drawing analogies between capitalism's parasitic appropriations of space and time and the behavior of viruses, such conclusions would be a categorical error that misses the point of such critique. There is nothing inevitable about fossil capitalism's ascendency during the modern era, nor about its constitutive blindness to the true costs environmental and labor exploitation. Likewise, though humans and our apelike ancestors have coevolved with viruses for millions of years this in no way dooms our species to parasitic relationships with each other or with more-than-human agents. Although the capitalist ideology at work in theories such as social Darwinism (examined in Chapter 4) may appropriate concepts from evolutionary science to naturalize its political claims for the sake of autocratic social order, my goal in exploring capitalism's viral entanglements is precisely the opposite: to illustrate how capitalism's compression of time and space aims to foreclose future uncertainty but in practice actually *increases* the risk of future destabilization in the form of chronic disease outbreaks. Or, as Malm puts it, because we live in a world where

> capital is fastened to ever more land and sucking its contents into circulation at an ever madder pace … this must, as a general law, result in a high risk of zoonotic pandemics, as one consequence of the ecological havoc caused. (81)

Neoliberalism's "parasitic" praxis is not a product of humanity's "hard-wired" biological evolution, but rather of socially contingent forces that could always be otherwise. Which is to say that Agent Smith's "revelation" in *The Matrix* that the human *species* as a whole is a "virus" destroying planet Earth actually misses the mark, for reasons I will examine in the next section.

Hormone Robots, Broken Clocks and Hothouse Flowers: Viral Capitalism in Atwood's MaddAddam Trilogy and Bacigalupi's *The Windup Girl*

In her article "What's the Matter with Dystopia?" Ursula K. Heise explores the popularity of apocalyptic fiction focusing on eco-disasters, totalitarianisms and pandemics, asserting that while work in this genre may "aspire to unsettle the status quo" most ends up "reconfirming it" by "failing to outline a persuasive alternative." Citing examples such as Atwood's *MaddAddam* (the third novel in Atwood's trilogy), Heise discusses how many eco-apocalyptic novels perpetuate clean-slate fantasies of wiping away late capitalist society and romanticize survivalism. Via a clichéd "attention to the routines of everyday life" that makes survivalists hard to tell apart from DIY hipsters, novels like *MaddAddam* end up being escapist and "comforting" rather than truly disturbing:

> What really counts is that the characters, in their break from the corruptions of the past, no longer have to deal with things like crowded cities, cumbersome democracies, and complex technologies. Whatever the hardships of their lives may be, they are better off without the world of corporations, biotech, and the Internet—even, apparently, at the price of genocide.

As readers familiar with Atwood's MaddAddam trilogy know, the novels depict the collapse of a future totalitarian society that, unlike that depicted in *The Handmaid's Tale*, is run by a capitalist technocracy rather than a fascistic theocracy. In "Corp(Se)ocracy: Marketing Death in Margaret Atwood's *Oryx and Crake* and *The Year of the Flood*," Sarah Appleton calls this government "Corpocracy" (64) to signify how it intensifies destructive aspects of current neoliberal capitalism, pushing its greed, neocolonialism, racism, sexism and speciesism to such grotesque extremes that readers are encouraged to understand (if not sympathize with) bioengineer Crake's motives for creating an artificial pandemic that wipes out this society.

I agree with Heise that the MaddAddam trilogy largely fails to depict a compelling post-apocalyptic world that represents a progressive alternative to Corpocracy, reflected in the fact that only a fraction of the trilogy actually attempts to imagine the new world that arises from the ashes of the old. Atwood's focus in the first two novels (and much of the third) is on satirizing the shortcomings of pre-apocalyptical society, a world similar to ours but whose failures go far beyond "crowded cities, cumbersome democracies, and complex technologies." I would argue that the technocratic dystopia in the trilogy is a truly disturbing vision of what I am calling "viral capitalism" insofar as it depicts hegemonic neoliberalism's potential to colonize both space and time, foreclosing any truly open relationship to the future. Among the most disturbing symptoms of this foreclosure is the absence of any meaningful sense of *history* in the novels, such that the past as well as the future's contingency disappears in a fog of presentism. Readers are presented with a nightmarish vision of capitalism ruthlessly reproducing itself by reducing all life to commodities, exploiting disempowered "others" (both human and more-than-human beings) and systematically denigrating the humanities in order to promote biological essentialism. While the trilogy is disappointing in its failure to depict alternatives to viral capitalism, I think it succeeds in showing hegemonic forces at work in Corpocracy that make the ability to imagine such futures virtually impossible for those living inside this system.

Atwood has resisted labelling the MaddAddam trilogy "dystopian," preferring the term "ustopian" because it better reflects her notion that "within each utopia [there is] a concealed dystopia" and (likewise) that "within each dystopia [there is] a hidden utopia, if only in the form of the world as it existed before the bad guys took over" ("Dire Cartographies" 85). However, readers of the trilogy may find it very difficult to locate a kernel of hope in the pre-apocalyptic world that exists before Crake's "waterless flood" pandemic. Indeed, as several critics have pointed out, Atwood breaks with the apocalyptic novel's convention of representing the pre-apocalyptic era as a "better" time recalled with nostalgic fondness. Instead, the pre-pandemic society represented in *Oryx and Crake* (2003) and *The Year of the Flood* (2009) is hellish, consisting of tightly controlled corporate compounds where the technological elite live and work, and chaotic, ecologically devastated and crime-ridden "pleeblands" where the majority of the world's human population ends up. In the absence of governmental regulation, unfettered fossil capitalism has colonized the globe and runaway climate change has destabilized weather and food production, along with increasing disease outbreaks. As in Malm's diagnosis of viral capitalism, Corpocracy's unsustainable practices, ranging from deforestation to produce GMO coffee to industrial meat production, fosters serial pandemics. Significantly, a character named Jimmy (one of *Oryx and Crake*'s central protagonists) relates that his "earliest complete memory" (15–16) is of a "huge bonfire" of cows, sheep and pigs: unbeknownst to his five-year-old self, these animals were probably being culled to stop a disease outbreak, such as the 2001 foot-and-mouth FMD epidemic in Britain (which resulted in the slaughter of 6 million cows and sheep) or the BSE "mad cow" outbreaks of the 1980s and 1990s, a zoonotic prion disease that affected cows and humans (with 4 million cattle being slaughtered as a containment measure). Although the novel doesn't specify what epidemic led to the animal bonfire Jimmy witnesses, the historical outbreaks that this event mirrors were both facilitated by feeding farm animals (pigs and cows, respectively) the remains of other infected animals, a cost-effective alternative to a vegetarian diet. In the absence of effective oversight, such outbreaks are common in *Oryx and Crake*'s hyper-neoliberal Corpocracy. At one point Jimmy reflects on the endless stream of repetitive disasters reported on news broadcasts: "more plagues, more famines, more floods, more insect or microbe or small-mammal outbreaks, more droughts, more chickenshit boy-soldier wars in distant countries. Why was everything so much like itself?" (253–4). Here Jimmy reflects both a sense of disaster fatigue common to people from relatively privileged backgrounds in this world, as well as ignorance of how the history of white supremacist, capitalist colonialism set the stage for Corpocracy's global hegemony and for violent social conflict/resistance among colonized peoples that he dismisses as "chickenshit boy-soldier wars" abroad.

In stark contrast to Jimmy's ahistoricism, the novel makes it clear from the outset that Corpocracy is an outgrowth capitalist colonialism. As one of the few survivors of Crake's engineered pandemic, a traumatized Jimmy struggles to think clearly in the opening chapter, remembering bits and pieces of things he may once have read in his humanities classes at Martha Graham Academy:

> "It is the strict adherence to daily routine that tends towards the maintenance of good morale and the preservation of sanity," he says out loud. He has the feeling he's quoting from a book, some obsolete, ponderous directive written in aid of European colonials running plantations of one kind or another … Rubber plantations, coffee plantations, Jute plantations. (What was jute?) They would have been told to wear solar topis, dress for dinner, refrain from raping the natives. It wouldn't have said *raping*. Refrain from fraternizing with the female inhabitants. Or, put some other way … He bets they didn't refrain, though. Nine times out of ten.
>
> (4–5)

Although Jimmy can't quite overcome his amnesia regarding the source of "directives," readers familiar with texts like *Heart of Darkness*—specifically Marlowe's meditations on the saving grace of *routine* as a discipline encouraging "restraint" among European colonizers of Africa who might otherwise succumb to the kind of "dark" passions that consume men like Kurtz—might experience an uncanny resemblance between colonial history and Corpocracy's ongoing environmental and social injustice. Nineteenth-century Euro-American "rape" of colonized places and peoples has been replaced by climate colonialism in Jimmy's world, corporate-led deforestation, the cultivation of GMOs to replace indigenous labor, debt economies and the trafficking of children as sex slaves under Corpocracy.

It is no accident that in this near-future world the elite class is (still) comprised mostly of white males whereas women and people of color make up most of the "plebian" underclass. The domination of these "othered" groups is perpetuated via their animalization, as is the case for a mysterious character named "Oryx," a woman from an unspecified Southeast Asian country whose "beautiful cat's face" (255) and poverty doom her to be sold into a life of child pornography, or impoverished US women like Toby and Ren in *The Year of the Flood* who are (similarly) linked to more-than-human entities when they work at "Scales and Tails" (a sex club in the pleeblands frequented by members of the techno elite). Importantly, Oryx's suffering stems from the interrelated forces of climate change, Corpocracy and patriarchy. Her parents are subsistence farmers who struggle to survive in a world where "the weather had become so strange [it] could no longer be predicted—too much rain or not enough, too much wind, too much heat—and the crops were suffering" (*Oryx and Crake* 118). When Oryx's father dies, her mother copes with this loss of financial support for her large family by making the heart-wrenching decision to sell her youngest daughter to a "businessman" named "Uncle En"; she's then trafficked to a series of other exploitative men once her "uncle" is murdered and eventually winds up being employed as a sex worker for the Watson–Crick Academy, before becoming Crake's lover/employee. Insofar as the novel illuminates Oryx's dual exploitation as a woman and a person of color from a colonized country, it can be read as a text with a strong ecofeminist message, as Javier Martin contends in his essay "Dystopia, Feminism and Phallologocentrism in Margaret Atwood's *Oryx and Crake*."

However, within Jimmy's hyper-corporate society, no one—not even members of the relatively privileged technological classes—escapes the reach of a commodity culture that reduces life (and death) to processes that generate financial gain. As Appleton notes, Corpocracy no longer makes its profits by marketing traditional goods and services but instead generates "enormous capital gains [that] … rely upon a carefully premeditated market of, and for, death, a death hidden in the very products of con sumption" (64). Examples of such production include pharmaceutical companies like HelthWyzer that make vitamins laced with bioengineered diseases which require consumers to buy HW medicines to cure themselves, ad infinitum. As Jimmy's technologically gifted friend and fellow compound member Crake explains,

> the best diseases from a business point of view … would be those that cause lingering illnesses. Ideally—that is, for maximum profit—the patient should either get well or die just before all his or her money runs out.
>
> (211)

Crake's father, a high-ranking tech developer for HelthWyzer who discovers the covert vitamin scheme and threatens to become a whistleblower, is killed in an act of corporate sabotage. Examples of such morally bankrupt profit-generating schemes underscore the extent to which viral capitalism in the novel is not merely a product of unanticipated "collateral" damage accompanying biologically unsound business practices, but is taken to another level insofar as pathogens are deliberately *engineered* by Corpocracy. Although Crake claims to admire the "really elegant delivery system[s]" (211) of such designer diseases, his father's murder contributes to his misanthropy and to his belief that human beings are constitutionally flawed, deserving genocide via a pathogen he later engineers with a similar covert delivery system (i.e. the virus is embedded in BlyssPluss pills marketed as libido enhancers and protection from STDs). Before creating the BlyssPluss pill, Crake tests prototypes of its pathogen on his own mother and his stepfather (Uncle Pete), murdering them because he believes they betrayed his father to HelthWyzer. When his mother becomes ill at work and has to be isolated in a quarantine area, Crake looks on with apparent dispassion from an observation window as her body dissolves.

Given scenes such as this, it would be tempting write off Crake as a recognizable stereotype: a neuroatypical male scientist (albeit one who dedicates himself to Hamlet-style revenge) who is incapable of empathy. Yet such a reading would detract from a more subtle understanding of Atwood's critique of viral capitalism as a force that shapes both Crake's *and Jimmy*'s empathy deficit by rewarding quantitative, objectifying and reductive thought, while denigrating the arts and imaginative structures that could foster more humane treatment of living beings in general and more historically informed understandings of human potential. Corpocracy has clearly sided with science in the "two cultures" debate given that technology is the engine of capitalist expansion into new markets, particularly those involving biotechnology. "Biodiversity" in this world is synonymous with the profitable production of transgenic organisms, living commodities that can be designed to meet a range of demands, all of which are unapologetically anthropocentric. Medical applications include hyper-intelligent "pigoons," transgenic pigs designed to grow organs for human transplant and to help stroke victims by re-growing human neocortex; aesthetic products such as "cute" GMO animals designed to be pets ("rakunks") and algae-derived "Smart Wallpaper" (201) whose color changes according to a human occupant's moods; and food applications, such as "ChickieNobs" (203), headless flesh bulbs grown as meat which can feel no pain, presumably in order to relieve human consumers of any guilt that might otherwise attend such consumption. Given the extreme alterations such patented lifeforms are subjected to, it is advantageous (from Corpocracy's perspective) to minimize any emotional response on the part of prospective producers or consumers of such commodities. Hence the empathy that Jimmy feels as a child towards his own duck-shaped boots at the farm animal bonfire and towards pigoons in his father's lab is largely weaned out of him via a steady diet of hyperviolent internet media as a teenager, including "animal snuff sites" and those featuring live human executions such as "headsoff.com" (82). Likewise, Crake cultivates a persona of cool, scientific detachment as an adaptation strategy because he knows that to succeed in Corpocracy he'll need to treat living things instrumentally, reducing them to engineering projects for which he will be rewarded financially and be held in high esteem. Hence, his emotional repression is far greater than Jimmy's who, as a mere "word person," is destined for life as a second-class citizen and, therefore, can indulge in contemplating useless emotions such as romantic love, empathy and grief. Indeed, Crake's belief that human beings are little more than biologically hard-wired "hormone robots" (166) whose desire for sex, status and material gratification hopelessly maladapt our species to environmental sustainability is less a scientifically accurate understanding of human evolution than it is a reflection of the ideal consumer under corpocracy. Yet because Crake's understanding of the humanities is impoverished—he ludicrously reduces art to an "amplifier" that male artists employ as a strategy for "getting laid" (168) and "imagination" to a death-denying "aphrodisiac" (120)—he cannot see how or why his biological essentialism is *ideological*, largely a product of social conditioning.

One of Corpocracy's major innovations from the standpoint of social control is to denigrate the humanities, treating such studies as fields for those who are not intelligent enough to pursue science and as mere "window-dressing" in the service of capital, something designed to decorate "the cold, hard, numerical real world in flossy 2-D verbiage" (188). A symptom of this society's cultural impoverishment is its profound *ahistoricity*, ways the past has been flattened into a series of meaningless events with little relevance to the lives of citizens. Systemic political reform is no longer possible in Corpocracy's autocratic security state, and the future is foreclosed into a repetition of the same capitalist dystopia (hence, Jimmy's aforementioned sense that time has become a fun-house maze in which every new catastrophe mirrors previous ones). The profound absence of historical perspective in the novel is signaled from the outset in the novel's opening chapter, wherein a post-pandemic Jimmy consults his watch "out of habit," only to be confronted by the "blank face" of its broken mechanism. Most commentators read Jimmy's uncanny "terror" (3) in this moment as a symptom of the fact that he hasn't yet adjusted to post-pandemic realities (i.e. clock time's absence in a world without civilization), but it would be just as apropos to read this as a lingering symptom of history's status in *pre*-pandemic Corpocracy. The clearest examples of history's attenuation in this world are online games that Crake and Jimmy play, such as "Blood and Roses," a trading game wherein human atrocities and achievements are assigned numerical values ("the number of corpses" for atrocities and "the latest open market price" for artistic achievements) and "the player who managed to retain the most human achievements by Time's Up was the winner" (79). By being reduced to numerical trading tokens, any atrocity can be played against any achievement regardless of historical incongruity ("one *Mona Lisa* equalled Bergen-Belsen, one Armenian genocide equalled the *Ninth Symphony* plus three Great Pyramids") and in complete disregard of any historical context that might make the achievement or atrocity meaningful. The game is, thus, a microcosm for the market's overdetermination in corpocracy: trade and "knowing the numbers" (79) enough to beat your competitor are all that ultimately matters. Likewise, this "wicked game" always favors the Blood side:

> That was the trouble with Blood and Roses: it was easier to remember the Blood stuff. The other trouble was that the Blood player usually won, but winning meant you inherited a wasteland. This was the point of the game, said Crake, when Jimmy complained.
>
> (80)

Ominously, the "take-away" message about human history that Crake gleans from playing this game is our species' proclivity for horrific violence, genocidal tendencies that are not counter-balanced by achievements in the arts and sciences but rather (he later concludes) are part of our faulty biological makeup. In the absence of historical perspective about factors that have contributed to atrocities, it is easy for Crake to conclude that humanity's destructive behavior is hard-wired in our species.

Another symptom of ahistoricity in Atwood's "ustopia" is the lack of efficacy that political protests and calls for change have in this world. Large-scale protests do, at times, erupt such as the Happicuppa coffee riots, a peasant- and small grower-led uprising against a Starbucks-esque coffee corporation that develops GMO beans which can be machine- (rather than hand-) harvested thereby replacing any need for human workers. Yet there's no evidence that such eruptions result in corporate reform or improved labor conditions. In the case of the Happicuppa protests, the uprising lasts "day after day" and involves riots, looting, kidnapping of Happicuppa personnel, a failed Boston "Coffee Party" and a bombing of the Lincoln Memorial, but at each juncture protests are violently put down by "soldiers" and police. Uncle Pete, who owns a lot of Happicuppa stock, is unmoved by the protests, commenting sardonically that this is nothing more than "the usual uproar" and that agitators will get tired of protesting because "[e]verybody wants a cheaper cup of coffee—you can't fight that" (180). Though Crake (ironically) describes the protests unfolding on television as "history in the making" (179) and thinks that the soldiers bludgeoning peasants should be "whacked," he clarifies that he thinks the Happicuppa corporation should be punished *not* because they are killing peasants, but because "they're nuking the cloud forests" to plant GMO crops. When Jimmy suggests the peasants would also be clearing the forests if they could, Crake agrees and concludes that there really "aren't any [human] sides" worth advocating for in the conflict. Thus, protests that could have been perceived by both teens as a clear example of a conflict created by environmental injustice (i.e. Happicuppa's parallel domination of "othered" peasants and biologically diverse rainforests) are rendered into an ahistorical "all sides are equally culpable" media spectacle. Furthermore, it is worth noting that this is the first hint that Crake's political impulses have a deep ecological dimension. His disregard for peasants killed in Happicuppa riots ("there's always been dead peasants") and genuine concern for rainforests being destroyed to grow GMO crops suggests that he cares more about loss of biologically diverse ecosystems than about human life. Indeed, his obsession with the game "Extinctathon" and consistent critique of human overpopulation underscore this point.

Just as human history has been reduced to a trading game in "Blood and Roses," the problem of mass extinction going on in the world outside of Corpocracy's high-tech sanctuaries is reduced to a trivia game in "Extinctathon." As I mentioned at the start of this chapter, contestants in this competition log into Extinctathon's website (monitored by the MaddAddam organization which names "dead" rather than living animals), and play with a MaddAddamite who provides clues regarding "some bioform that had kakked out within the past fifty years" (80), which contestants use to guess the animal's identity. Crake claims to be obsessed with this game "because [he's] good at it" and Jimmy assumes that his geeky friend wants to become a Grandmaster of the game "not because it meant anything but just because it was there" (81), yet the fact that Crake replaces his given name "Glenn" with the codename he uses to play this game suggests that it has a deeper significance for him. The red-necked crake was a relatively rare, elusive Australian bird before it went extinct under Corpocracy, and oryxes were elegant, graceful antelopes. In addition to having personal affinity with their codename animals, it is likely that Oryx and Crake take these names instead of their given ones as a way of memorializing these lost species, remembering and (on an unconscious level) grieving their loss. In *MaddAddam* we learn that Crake has ties with bioterrorists who use the Extinctathon game as a front, including Adam One (leader of the God's Gardener's cult) and an ex-biologist named Pilar, both of whom give a teenage Crake disease microbes smuggled out of HelthWyzer which he later develops into "the core of the BlyssPluss pills" (330). The God's Gardener religion hybridizes aspects of Judeo-Christian/Eastern theology and ecological science in order to create a holistic faith based upon principles of bio-egalitarianism. It also combines concern for loss of biodiversity with a conviction that "the human race was overdue for a population crash"—all of which are elements of deep ecological thought. Indeed, as Gerry Canavan notes in "Hope, But Not for Us," Crake's motives for genocide may not only have been a sense that humanity is congenitally flawed and "cannot learn" to change (*MaddAddam* 291) but also a desire to save the Earth's remaining species before they are wiped out:

> So-called 'deep ecology,' both in and outside science fiction, has long wrestled with … this fraught relationship with catastrophe and extinction in its push for a human race with so light an ecological footprint so as to … be erased from the planet altogether … In the face of man's mania for self-destruction, which now threatens not only the future of *Homo sapiens* but all life on the planet as a whole, Crake's virulent misanthropy unexpectedly becomes transformed into something like a virtue, the truest form of love … genocide as environmental policy: global death, in the name of preserving life.
>
> (150)

As should be apparent from my analysis, although Crake may (in his delusion) believe his genocidal pandemic is a "loving" act that will preserve more-than-human biodiversity, it is (in fact) the logical culmination of what I have been calling "viral capitalism." Like the Corpocracy he hates, Crake uses coercive methods to further his own ends, secretly embedding pathogens in pharmaceuticals marketed as life-enhancing drugs. Like corpocracy's elites, he feels entitled to make the choice of who gets to live or die because he deems the general public to be "hormone robots" incapable of rational thought. He may justify his biological essentialism as a viewpoint confirmed by evolutionary psychology, but without viable alternatives to scientific accounts his view of humanity is fatally reductive. Without an informed knowledge of human history, he is doomed to repeat the mistakes of previous fascist, genocidal leaders and of the autocratic economic regime he lives under. For many readers, a question that looms over the entire trilogy is whether Atwood succeeds in exposing (and critiquing) Crake's biological essentialism or whether she herself embraces such views. A survey of the trilogy's critical reception suggests that scholars are deeply divided over this issue.[3] For me, Atwood's personal allegiances on the question of biological determinism are secondary to what I believe the trilogy demonstrably accomplishes: namely, a chilling depiction of viral capitalism's capacity not only to destroy biodiversity, but also to colonize time to such a degree that human history becomes extinct and is misrecognized as "human nature."

Although Paolo Bacigalupi's *The Windup Girl* diagnoses some of the same problems arising from viral capitalism as those addressed in Atwood's MaddAddam trilogy (including the creation of GMOs for profit regardless of ethical concerns) there are also important differences in this novel's representation of human and more-than-human subjects, as well as its engagements with temporality. Bacigalupi's depiction of a post-fossil-fuel Thailand is far more suspicious of genetically determined identity than is the case in Atwood's trilogy and (consequently) both humanoid and viral agents behave in non-deterministic ways in *The Windup Girl*, suggesting a post-apocalyptic future that is radically open and will not hinge upon bioengineered "solutions" to capitalist inequities. Likewise, this text's representation of fossil capitalism's long-term climate effects even in a world powered by kinetic energy as well as its highlighting of genetic engineering's impacts upon evolution make viral capitalism's geological-scale agency visible in compelling ways. Though Bacigalupi shares Atwood's sense that neoliberal capitalism is driving the world to ecological collapse, his novel deconstructs humanist essentialism more fundamentally than Atwood's trilogy and (in doing so) creates opportunities for thinking emergent queer kinship networks that may arise in a future that is not foreclosed by capitalist temporality.

The Windup Girl's setting in a post-"Expansion," post-fossil-fuel Thailand is significant as a move that decenters a focus on how neoliberalism affects the global north to one that explores how this system impacts those in the global south—particularly in countries with a strong nationalist identity that resists foreign intervention. The novel's dystopian Bangkok society arises in the aftermath of fossil capitalism at its zenith, exploring a period of contraction in which the quest for viable food commodities ("calories") becomes the dominant economic driver in the wake of agricultural disasters created by climate change and plant viruses engineered by biotech corporations attempting to corner the market on GMO crops. Significantly, although petromodernity has long ceased being capitalism's primary structural manifestation during this time, neoliberal colonialism, environmental destruction and social inequities continue unabated under a new kinetic energy regime. Thus, in its diagnosis of the Anthropocene the novel seems closer to "Columbian Exchange" (or "Plantationocene") theories that locate the "Capitalocene"'s origins in early modern New World colonialism rather than accounts which locate the starting point in the rise of fossil capitalism during the late eighteenth and early nineteenth centuries. In twenty-third-century Thailand, what Vandana Shiva would term "biopiracy" (rather than energy extraction) has become the key to generating wealth, and large biotech companies with headquarters in countries like the US engage in corporate espionage to locate seedbanks abroad, archives of genetic information that can by stolen and "generipped" to produce new GMO crops that are (in turn) introduced as monocultures to replace indigenous food sources. As Shiva explains in *Biopiracy: The Plunder of Nature and Knowledge* and elsewhere, transnational biotech firms such as RiceTec (a Texas-based corporation) routinely appropriate genetic material from crops cultivated over long periods of time in the Global South to "create" GMO crops that can be patented for profit, a form of theft that aims at "privatizing the generative capacities of life, common spaces and the knowledge produced by people over time" ("Sexual Economics" 67). *The Windup Girl* opens with American "calorie man" (corporate spy) Anderson Lake marveling at a flavorful ngaw fruit he happens upon in a Thai market, an organism that "shouldn't exist" in a sea of bland GMOs and that, unlike such produce, is mysteriously resistant to "the agricultural plagues of the calorie companies" (2). Lake suspects that the gnaw has been resurrected from a hidden Thai seedbank of non-genetically engineered plants and becomes obsessed with locating and stealing this treasure even as he can't help but admire the foresight of his would-be victims, noting "The Thai Kingdom is clever where others are not. It thrives while countries like India and Burma and Vietnam fall like dominoes, starving and begging for the scientific advances of the calorie monopolies" (3).

Despite Lake's elaborate schemes to access the seedbanks by manipulating rival members of the Thai government (siding with its open market Trade Ministry over its protectionist Environmental Ministry), his biopiracy is ultimately undone by more-than-human agents' capacity to behave unpredictably. As in Atwood's MaddAddam trilogy, corporations try to weaponize disease vectors to further their ends in *The Windup Girl*, but in Bacigalupi's novel they cannot control what they unleash via their greed-driven practices. As aforementioned, in *Oryx and Crake* pharmaceutical companies embed diseases in their drugs in order to create illness and captive consumers, and Crake replicates this process when designing his BlyssPluss pills. With notable exceptions like the transgenic Crakers or relatively minor products like "rockulators" (rocks designed to store water that sometimes explode) most of the technology represented in the trilogy does what it is designed to do. In contrast, disease agency is radicalized in *The Windup Girl*'s depiction of viral capitalism as wave after wave of pandemics cross plant–animal species boundaries in unforeseen ways. The novel makes it clear at the outset that GMO crops and bioengineered plant infections have spontaneously mutated to jeopardize agriculture around the world and indicates that some of the latter have also developed the ability to infect animal and human populations. As Heather I. Sullivan argues in her essay "Cross-Infections of Vegetal–Human Bodies in Science Fiction," diseases catalogued in the novel such as "blister rust," "Nippon genehack weevil" and "cibiscosis" deform the infected with horrific symptoms, offering up "not only the spectacle of contaminated bodies" but also "shed[ding] light on both the life-giving and dangerously trans-corporeal aspects of our living entanglements and ecological dependencies" (345). It is precisely such ecological entanglements that viral capitalist corporations depicted in *The Windup Girl* continue to disavow in their quest for the next big opportunity for economic "expansion." As Andrew Hageman observes, calorie men like Lake who work for corporations such as AgriGen dream of "grow[ing] a superior version of the same capitalist economic model that led to … mass decimation in the first place" (289) and inadvertently catalyze new diseases via unsound ecological and business practices. Algae in a kink-spring factory that serves as a cover story for why Lake is in Thailand unexpectedly mutates due to lax safety standards, producing a disease that eventually attacks the calorie-man's lungs and leads to his demise.

Lake's corporate plans are undone not only by such disease agency but also by more anthropomorphic products of biotechnology such as Emiko, the titular "windup girl." Engineered to resemble a doll-like automaton with "stop-shutter" movement whose prime directive is to be subservient (sexually and otherwise) to her masters, Emiko evolves over the course of the narrative into a powerful agent and potent symbol for the dual exploitation of women and nature under viral capitalism. Like *Oryx* in Atwood's trilogy, Emiko is a beautiful Asian who falls on hard times when her protector (a Japanese businessman) abandons her to a life of poverty and prostitution in Thailand. However, as a bioengineered "new person" whose every movement emphasizes her artificiality, Emiko's abuse at the hands of unscrupulous human beings is of a different order than Oryx's. She is not only trafficked in a Thai brothel but is subjected to sexual abuse in "shows" designed to entertain patrons, typically "small men who imagine themselves to be at the vanguard of some new Expansion" (36). Emiko's ritualized "humiliation" as an "almost human" (34) who is treated like an "animal" (38) is a way of underscoring her ontological difference (and inferiority) to human beings; yet her abuse also mirrors women's more general dehumanization under patriarchy, the historical plight of women of color under colonialism, and the mistreatment of animals under industrial capitalism. Adding to Emiko's suffering is the fact that her pores are too small to regulate her body temperature effectively in Thailand's sweltering, climate-change-induced heat. Though this feature makes her skin particularly smooth to touch, it is only practical in air-conditioned, indoor environments. As Lake observes, Emiko's "model must only sell to elites" who would "have climate control" (44); such engineered weakness would be "worth the trade-off, to them" but severely compromises Emiko's health. The calorie-man becomes sexually involved with Emiko because he thinks she represents "the opposite of the invasive plagues that he fights every day," seeing her as "[a] hothouse flower dropped into a world too harsh for her delicate heritage" who will likely not "survive for long"

(61). It is Emiko's perceived "vulnerability" that attracts Lake, moving him to pity her "pretended strength when she had nothing at all."

As it turns out, Lake is wrong on all of these counts. Emiko proves to be very much *like* the "invasive plagues" that calorie-men under viral capitalism both create and battle when they prove difficult to control, only unlike her human counterparts she is resistant to infection from such diseases and (in this sense) is more resilient than Lake in the long run. Likewise, she shows extraordinary strength in breaking from her genetic and social programming to lash out against those perpetuating violence against her. Following Emiko's brutal gang rape in front of the Somdet Chaopraya (Crown Protector of the Thai Royalty)—a crime that results from Lake's attempts to attain leverage over this official by catering to his sexual sadism—Emiko kills all of her assailants, including the Crown Protector. Lake's calculated betrayal of Emiko suggests that, as with the ngaw fruit he covets, he desires her as an exotic commodity to be temporarily savored but also readily sacrificed in his quest for power. His total failure to anticipate Emiko's agency and capacity for violence parallels his failure to anticipate the "thing-power" (in Jane Bennett's parlance) of genetically modified plants and pathogens that, despite the efforts of biotech corporations, often mutate in ways that result in unforeseen human suffering.

Indeed, it could be argued that agents of viral capitalism like Lake only desire beings like Emiko if they can rely on such entities to be ephemeral in the sense of having a kind of built-in biological obsolescence. As a "new person," Emiko is designed to be infertile and (as aforementioned) Lake assumes that it is "unlikely that she will survive for long" in her harsh circumstances. Insofar as her capacity to reproduce has been limited by design, Emiko resembles GMO "terminator seeds" designed to be fertile for only one season so that farmers are compelled to buy such seeds annually. Shiva links such practices with "biopiracy," but of a kind that aims at "a kind of colonialism of the future." As she explains in a 1999 interview published in *Atlantis*,

> The capacity to patent life basically implies that today corporations can start extracting surplus from the future—the future of humanity and the future of nature… This is being achieved by things like the "terminator seed" technology where after the seed is planted there is a switching on of the genes that prevent germination. Nothing could be more dramatic than an appropriation of surplus value from the future in the today by saying "no reproduction without my permission"—without the permission of capital in the hands of white patriarchy.
>
> (71)

At its most fundamental level, *The Windup Girl* explores exactly this kind of struggle for who gets to control not only food and energy production but also, more profoundly, *reproductive* power: the "seeds" of the future.

From this perspective, the fact that Emiko is one of the few survivors of a both an algae pandemic and a flood that destroys Bangkok in the novel's conclusion is a testament to her resilience, a capacity to endure into the future that is augmented when she discovers she may (in fact) be able to pass on her genes to another generation. Having taken up residence in an abandoned apartment in what was downtown Bangkok before the flood, Emiko is visited by Gibbons, a bioengineer, and his trans partner. Ill with an unspecified disease, Gibbons admires what he sees as Emiko's superiority to human beings (including her resistance to bioengineered pathogens) and anticipates a future in which "all people will be New People," implying he may be able to create fertile clones using Emiko's genetic material. Scott Selisker reads the novel's ending as a meditation on ways that "genetic modification will count as a portion of humankind's geological agency, as the genes we develop will spread in time and space beyond the confines of humanity as a species" (514). That is, although corporations in the novel such as AgriGen create GMOs in an attempt to colonize the future for capital gain, what they create will (ironically) not only elude control but will also likely establish new, unanticipated evolutionary pathways. Thus, Selisker interprets the conclusion as a vision of time liberated from capital that heralds the rise of a new genetic "wilderness" that "encourages us to consider the fate of genetic materials over the *longue durée* of an unforeseeable future." Although this vision may initially seem similar to the fate of Atwood's transgenic Crakers in that they are also beings well adapted to a climate-altered, post-pandemic world who develop capacities that exceed their design, as a speculation about the future Bacigalupi's novel ends on a far more post-humanist note. Atwood's post-pandemic humans hope that something of their species will survive via biological reproduction (i.e. in their capacity to mate with the Crakers) and in the Crakers' "essentially human" capacity to create art; whereas Bacigalupi's ending imagines a future race of cloned "New People" who will share virtually no biological or cultural kinship with their human creators. As if to underscore the fact that *Homo sapiens* will be an evolutionary dead-end, Emiko rejects Gibbons's hubris in continuing to think that he has god-like power over GMOs and registers her disgust with humanity in general, declaring "I hate your kind" (358). Symbolically, agents of viral capitalism like Lake and Gibbons are doomed to succumb to the diseases they have created, though they may be succeeded by new entities capable of thriving even in capitalism's ruins. In radically breaking with any notion that biological reproduction is necessary for survival, the novel imagines new forms of "queer diversity" (300) may arise to replace natural evolution, as critics such as Hageman have suggested. This is a future that may hold hope for resilient GMOs like Emiko but is bleak for humanity. In the aftermath of a 2011 flood that devastated large parts of Bangkok, the 2020 coronavirus pandemic and predictions that much of Bangkok will be underwater by 2030 if climate change continues unabated, *The Windup Girl*'s critique of viral capitalism's disastrous socio-ecological effects grows more relevant each year.

Arboreal Entanglement in Deep Time: Richard Powers' *The Overstory*

"Trees stand at the heart of ecology, and they must come to stand at the heart of human politics" (454). This proclamation, made during a lecture on the theme "What is the single best thing a person can do for tomorrow's world?" by dendrologist Patricia Westerford in Richard Powers's *The Overstory* (2018), could serve as a summation of the novel's central aspiration. Most of the text's human characters do their part in trying to bring about progressive change, either via innovative science and technology (Westerford, Neelay Mehta), visionary art (Nicholas Hoel), revisions to the law (Ray Brinkman), or via eco-activism (Hoel, Douglas Pavlicek, Mimi Ma, Adam Appich and Olivia Vandergriff). Yet, despite such monumental efforts, it is less than encouraging that Patricia's aforementioned lecture may well be her last, as the novel suggests her answer to the question of what the "best thing people can do" for the future may be to commit suicide. In one of the text's branching timelines, "plant Patty," the character who knows the most about trees and their vital role in the world's ecosystems, kills herself in front of her audience by drinking toxic extract from an Angel's Trumpet (*Tachigali versicolor*) or "suicide tree." A reading that focuses on the novel's pessimistic elements would (similarly) note that by its conclusion all but two of the eco-activists are either dead or in jail and the one who is punished the most for his actions (Adam Appich) concludes that he and his deep green comrades-in-arms ultimately "accomplished nothing … *Not one thing*" via their monkey-wrenching (431, emphasis in original). In a novel of proliferating trees, Adam is also the only one of its nine human protagonists to have a child but he will never have the chance to raise his son given his 140 year jail sentence. In keeping with Nick Hoel's late revelation that it's human beings, not trees, that are "the most wondrous products of four billion years of life [that] need help" (493), it would appear that the novel prophesizes that our species is heading for a future "cataclysm" (481) that can't come soon enough for the Earth's remaining arboreal denizens.

And yet—but *still*. Such a tragic interpretation of *The Overstory* does not quite do justice to the complexity of what the novel accomplishes. For the purposes of this chapter, I will focus on two significant areas of achievement: first, ways the novel dramatically illustrates long-standing human and arboreal entanglements that challenge readers to generate a politics that does justice to this condition; and second, ways that it reveals the deep temporal dimensions of this entanglement such that time's openness (radical indeterminacy) is liberated from capitalist foreclosure. *The Overstory* is a novel that's fascinated by "branching": unpredictable, fractal patterns that characterize tree growth but also the proliferation of digital systems which obsess the visionary computer engineer Neelay and (later) his AI progeny. Pre-eminently, time is a fractal force in this text, apt to branch in ways that cannot be predicted in advance. Tragedy often strikes unexpectedly in the human characters' lives (as when Nick discovers his extended family dies on Christmas eve from gas asphyxiation or when Mimi's father commits suicide at a moment when his life's work comes to fruition), but just as often time swerves unpredictably in the direction of comic survival (as when Doug is caught by a Bo fig tree when parachuting from a plane during the Vietnam war, or Mimi avoids spending the rest of her life in jail because of Doug's negotiations with government authorities). Similarly, Powers commits authorial heresy by keeping the plot ambiguously open with regard to Patricia's "suicide," noting that when she lifts the poison-filled glass to her lips

> the world splits. Down one branch, she lifts the glass to her lips, toasts the room—*To Tachigali versicolor*—and drinks. Down another branch, this one, she shouts, "Here's to unsuicide" and flings the cup of swirling green over the gasping audience before exiting the room, leaving them to "stare at an empty stage".
>
> (466)

Although the specification of "*this*" branch could be read as suggesting that Patty doesn't take her own life in the novel's "real world" timeline (as opposed to one in a parallel universe), the fact that she makes no appearance in the novel's final "Seed" section but is accessible only via audio recordings and books that the other characters consume complicates this interpretation. On this plot point, Powers leaves readers deliberately suspended in uncertainty, a move which many may find unsatisfying but—as a commentary on time's openness—is nonetheless deeply meaningful. Arguably, in sacrificing knowledge of a major protagonist's ultimate fate the author is giving us an example of ways that our ontological being in the wider world can trump a fixation on "character," such that we resist "confus[ing] a satisfying story with a meaningful one" (383). In the following discussion, I will explore ways the novel brings into productive tension a paradigmatically tragic, neo-romantic view of our species' future (one wherein arboreal and temporal sublimity *minimizes* the significance of political action) and, in contrast, another which remains hopeful that one day a change in consciousness regarding multi-species justice may—unexpectedly—bring about a politics that honors our ongoing entanglement with trees and other more-than-human lifeforms.

A primary way that *The Overstory* defamiliarizes trees so that readers might learn to see them anew is by not only showing their influence on the "mayfly"-like duration of individual human lives, but also (in some cases) their longer, intergenerational history in some of the protagonists' family trees, and, at another level, their almost unimaginatively long, species-level agency in the tangled tree of evolutionary life on our planet. As critics such as Rachel Adams have observed, "at the level of content, each of nine human plotlines converges with the stories of trees" and these narratives are (in turn) "framed by the 'overstory' of the biosphere, with its botanical scales of time and space, subjectivity, and communication" (801). The entangled human–arboreal plotlines begin in the biographical "roots" section (wherein different tree species are paired with each human protagonist), and later these plotlines merge in the "trunk" section when five of these protagonists (Hoel, Ma, Pavlicek, Appich and Vandergriff) come together to engage in activism to protect forests.

In the case of characters such as Nicholas Hoel, six generations of the Hoel family are shown to have had a long-standing, entangled relationship with a single chestnut tree. This begins when Jørgen Hoel, a Norwegian immigrant who feasts on this tree's nuts in the eastern US, takes a few seeds west and plants them on his Iowan farm; eventually, one lone "sentinel" tree comes to maturity having survived a blight that wiped out others of its kind back east. Hoel's mid-western descendants begin to photograph this tree, creating a visual archive that records each month of its life until "extinction sneaks up" on both the tree and the Hoel clan—the former eventually succumbing to blight and the latter suffering the fate of other independent farmers in the age of Big Agriculture (i.e. taking on crippling debt to avoid being "swallowed up into [...] massive, managed, relentlessly productive monocrop factories" (17)). Following the aforementioned final extinction of the Hoel clan one Christmas eve, Nicholas is the last standing "watchman" of his kin, his fate mirroring that of the Hoel sentinel chestnut but also, as an artist, reflecting the fact he will function as a witness and (later) guardian of the future when he becomes involved in eco-activism and the creation of artwork designed to promote public concern about tree welfare. Punctuating the Hoels' human-scale generational history are glimpses of deep arboreal time—a radical temporal incommensurability between the scale of life as it's experienced by chestnut trees and individual members of the Hoel clan. Thus, as a dying Jørgen Hoel wonders "what makes the bark [of his chestnut] twist and swirl so, in a tree so straight and wide," we are told that

> [s]even hundred years before, a chestnut in Sicily two hundred feet around sheltered a Spanish queen and her hundred mounted knights from a raging storm. That tree will outlive, by a hundred years and more, the man who has never heard of it.
>
> (10)

Indeed, the opening Hoel history chapter demonstrates the extent to which the common perception of trees as static ("just there") in occidental culture is an illusion produced by these incommensurate temporal frames, an inability of short-lived animals to perceive subtle changes in our long-lived arboreal cousins. Only when Nick thumbs through 75 years of family photographs of the Hoel Chestnut in a "five-second" (film-like) succession of images does the tree's "secret. animated meaning" reveal itself, along with the way this tree "encodes" countless "gatherings in each season" (17–18) of the Hoel family. In this way, the novel frames its human stories within the "overstory of the biosphere" as Adams puts it; moreover, she asserts that "at all [these] levels, we encounter questions of care within and across species" (801).

While such issues of care frequently shape human–arboreal relationships depicted in the novel—encompassing both the joys and (in the case of the Covid pandemic which informs Adams's reading) the hazards of "webs of interdependency that have put the world in grave danger and that gesture to an uncertain future," Adams does not address ways that the novel's invocations of deep time often serve as a spiritual salve in the face of praxis that fails to honor such "interdependency." Though she identifies how the Earth's ecosystems and transnational capitalism function as "competing biopolitical forces so immense they cannot be fully apprehended by the traditional novelistic form" and (therefore) serve as factors that drive *The Overstory*'s structural innovations, she also claims that these "hyperobjects" (along with "viral networks" exposed during the Covid pandemic), "[do] not make human activity, or the stories of individual persons, irrelevant" (804) despite the systems-level perspective often adopted in Powers's novel. Yet the novel's depiction of human–arboreal relationships across deep time often exceeds an ethic of interpersonal care; indeed, such long-term entanglements often function as psychological compensation for our failure to treat trees with the care they deserve. Like the "geological" sublime I addressed in Chapter 1, Powers often evokes an arboreal sublimity in order to provide a broader context for individual traumas, to question the staying power of our long-term (collective) human impacts on the planet or to suggest forms of "kinship" that are impersonal or even unconscious.

A telling instance of this occurs when Patricia Westerford stops along the side of a highway in Utah to look at a stand of aspens, a "great, joined, single clonal creature that looks like a forest [and] has been around for the better part of a million years," making it "one of the oldest, largest living things on earth" (131). Unbeknownst to Patty, aspens like the stand she contemplates are not only entangled colonies that gather multi-generations of trees, they also entangle the novel's human protagonists. For the first time in the narrative, Powers brings together all the characters introduced so far in paragraphs describing their interactions with aspens: we are told that Mimi Ma and her family camped near this particular stand five years earlier; Nick Hoel "walks past the single quaking aspen in all of Central Park" twice over the span of 30 years; Doug Pavlicek makes "corral siding out of aspen" (132) for a fellow vet in Idaho; two aspens grow on lawyer Ray Brinkman's and his wife Dorothy Cazaly's property in St. Paul and are misidentified by Brinkman as "birches"; and Neelay Mehta and his father create digital aspens for a game they are developing that simulates walking through a "forest primeval." At this point in the novel "these people are nothing to Plant-Patty" *or to each other*, and yet their lives are said to "have long been connected, deep underground." Rather than being linked by interpersonal care, their kinship is said to "work like an unfolding book" at this meta-narrative moment, wherein disparate individuals and time frames are woven together via subterranean interconnections, such that their "past" (roots section) will only become "clearer" in a "future" that has not yet transpired. As if to illustrate this strange loop that encompasses both the past and the future, Powers cites a passage in a book called *The Secret Forest* that Patty has not yet written, yet is clearly inspired by her encounter with the aspen stand:

> You and the tree in your backyard come from a common ancestor. A billion and a half years ago, the two of you parted ways. But even now, after an immense journey in separate directions, that tree and you still share a quarter of your genes …
>
> (132)

In evolution's tangled tree, human DNA is entangled with that of our arboreal kin, just as (we have seen in the previous section) it is also entangled with viruses. While our distant ape-like ancestors adapted to life in trees, in the climate-change era trees like aspens are adapting to anthropogenic weather patterns by slowly "migrating" further north "where the ice sheets were" (133) in an attempt to survive.

When Patricia unexpectedly sees a condominium development "carved out from the heart of the spreading [aspen] clone" this abruptly punctures her deep time reverie, and she recoils as if receiving a blow: the sight "punches her in the chest." Capitalism in the form of vacation condos has already "cut through several acres of the root system of one of the earth's most lavish things." Although the luxury home market is literally erasing deep time in removing the aspen forest, Patty feels powerless to stop this destruction. Instead, in the aspens' "war" with human beings for natural resources, she assumes the trees will fall and "can hear, louder than the quaking leaves, which side will lose by winning." Similarly, when she later witnesses televised footage of police swabbing pepper spray in the eyes of eco-activists protesting the logging of old growth forests, her response is that such people are "beautiful" precisely because their "hopelessness makes them [more] determined" (304). Having previously testified about the ecological importance of old growth forests in a court case over proposed logging of such areas on public land in western Oregon, Patty knows what it means to "lose by winning": after the presiding judge places a hold on clear-cutting until environmental impact assessments can be done, she is told by the opposing side that all her advocacy has accomplished is to "make lumber a whole lot more expensive" (285) and has (thereby) incentivized accelerated logging on *privately* owned land. The lesson Patricia learns from such experiences is not that neoliberal capitalism is a fundamentally flawed system from an environmental justice perspective but (rather) that *Homo sapiens* as a species is fundamentally flawed, "doomed" to self-destruct. Thus, televised police brutality towards eco-protestors prompts her to ask her partner Dennis:

> How is extraction ever going to stop? It can't even slow down. The only thing we [humans] know how to do is grow. Grow harder; grow faster. More than last year. Growth, all the way up to the cliff and over. No other possibility.
>
> (304)

Patty's conclusion that there is "no other" possible path for humans other than destructive (and perpetual) "growth" resembles Crake's biological determinism in the MaddAddam trilogy insofar as both scientists conclude that our species is analogous to a cancer (or a virus) on the planet. Her honest answer to the question "*[w]hat is the single best thing a person can do for tomorrow's world?*" is, "from the standpoint of trees" (456, emphasis in original), to remove this growth: i.e. to commit suicide or, short of this, refrain from having children. Significantly, thoughts of deep time are not far from her mind immediately before she contemplates taking her own life before a horrified audience at a public lecture. "*Maybe*," she thinks, "*we want to hurt trees so much because they live so much longer than we do*" (466, emphasis in original). In Dr. Westerford's fundamentally tragic outlook, deep time and tree longevity are sublime phenomena that dwarf the scope of human lives and we try to compensate for this by destroying what makes us feel insignificant.

Readers of *The Overstory* might object that Patricia's political fatalism should not be conflated with Powers's outlook, given that hers is only one of nine perspectives explored in the novel. Yet none of the text's five activist characters manages to achieve much through their efforts. Olivia Vandergriff and Nicholas Hoel don't keep the giant redwood "Mimas" from being cut down despite months of tree-sitting in this tree. Mimi Ma and Douglas Pavlicek don't prevent a stand of Ponderosa Pines from being cleared near Mimi's workplace despite joining protests to keep this from happening. Adam Appich is no more effective than his co-conspirators when all five engage in "monkey-wrenching"—i.e. arson and property destruction after becoming exasperated by the inefficacy of legal protest. By the end of the novel, Olivia has died due to an accident during one of the group's covert operations, both Doug and Adam are in jail, Mimi is still free but wracked by guilt, and Nick has turned to artistic expression as an alternative to direct political action. In the end, none of these characters seems to believe that any of their activism "could make up for [Olivia's] death" (to quote Doug), and, as aforementioned, Adam asserts that they "accomplished nothing" (431).

However, it is instructive to contrast the futility of these characters' fictional activism with what real-life forest defenders who spearheaded the "Redwood Summer" in 1990 achieved over the course of 20 years of protests. Activists such as Julia Butterfly Hill occupied a 1,000-year-old redwood named "Luna" for over 700 days between 1997 and 1999 and kept this tree from being cut down (in part by raising funds to buy the tree from the Pacific Lumber Company); and, similarly, many redwoods marked for harvesting with spray paint were never cut down due to activism which drew the world's attention to the problem of harvesting old growth forests. The contrast between what was accomplished historically and what is represented in *The Overstory* suggests that, as Matthew Benjamin Cole puts it, "the political narrative of the novel is essentially tragic, showing the impasse at which existing activist and reformist strategies arrive" (142–3). Although Cole suggests that other, less "fatalistic" apocalyptic futures are also seeded by book's conclusion, these "emancipatory" (143) alternatives are, arguably, technological (Neelay's hope in his AI "learners") and spiritual (Mimi's Buddhist-influenced hope that a new world might arise "[o]nce *the real world ends*" (500), emphasis in original) rather than *political* in nature.

I would assert that the novel's evocations of deep time constitute a form of arboreal sublimity that attempts to compensate for this "essentially tragic" assessment of political action. Other critics, most notably Berthold Schoene, have analyzed sublimity's ecological importance in *The Overstory*. Drawing upon Elizabeth Carolyn Miller's concept of "dendrography" as "a form of ecological realism that strives to represent the natural world more accurately by inhabiting the scale and perspective of the arboreal" (Miller 711), Schoene unpacks the implications of what this means for human–arboreal communication in Powers's novel. In "Arborealism, or Do Novels Do Trees?" Schoene claims that *The Overstory*'s "dendrographic narration [is] not just *of*, but indeed *by* trees," and points to instances in the text wherein the narrative "voice" is "the inaudible, as-yet-unheard-of eloquence of trees" (1452) in passages that signal this by using special italic font. Schoene clarifies that "[w]hat is ultimately at issue here is the arboreal as a manifestation of the sublime" (1451), something he claims has a potential to "transform" characters such as Mimi and which "differs from its Romantic counterpart in that awe and humility in confrontation with the more-than-human do not subsequently translate into what Christopher Hitt terms 'self-apotheosis.'" That is, Schoene concurs with Hitt's interpretation of the "ecological sublime" as an experience with more-than-human otherness that has the power to "jolt us momentarily out of a perspective constructed by reason and language, a perspective that, in modern Western culture, has rendered nature mute" (Hitt 617). I have published extensively about the paradoxes entailed with attempts to interpret the sublime as a discourse that gives voice/meaning to inexpressible non-human otherness,[4] so will not rehearse my reservations about Schoene's claim that Powers's novel can legitimately be read as text *by* (not just about) trees. Instead, I would like to focus on the effects of the novel's evocations of sublime arboreal temporality as something which contrasts with the relatively short duration of human lives.

Although Laura DeLuca doesn't address sublimity, she explores the stark difference between human and tree temporal frames in her article "Temporality in *The Overstory* by Richard Powers," asserting that anthropocentric "quick-time" in the novel is "expressed [as] a parasitic dynamic of human life on earth" (80) that leads to environmental destruction. Likewise, in "Dendrography and Ecological Realism," Miller explores the gap between human and tree time in her reading of Thomas Hardy's *Under the Greenwood Tree*, asserting that the dendrographic perspective "is wider, in the sense of obtaining a greater height and distance from the earth, as well as longer, in the sense of temporal duration" and that "[a]n arboreal scale can thus achieve a certain *distance* beyond the individuated human life at the center of most forms of nineteenth-century realism" (711, my emphasis). By simulating the trees' "sublimely long timescale[s]" (697) and heightened view on earthly dramas, Hardy's dendrography "offers the tree as a symbolic model for thinking beyond the limited perspective scale of the human" (712).

When one studies numerous examples of "dendrographic" time in *The Overstory* and how this diverges from "limited [human] perspective scale," a striking commonality is the role that human *trauma* often plays in moments that address this contrast. An early, representative instance of this occurs immediately after Nick Hoel discovers his extended family has perished from gas exposure one Christmas eve. Having narrowly escaped death by stumbling out of the house and falling on the snow, he revives beneath the Hoel chestnut:

> When he looks up, it's into the branches of the sentinel tree, lone, huge, fractal, and bare against the drifts, lifting its lower limbs and struggling its ample globe. All its profligate twigs click in the breeze as if this moment, too, so insignificant, so transitory, will be written into its rings and prayed over by branches that wave their semaphores against the bluest of midwestern winter skies.
>
> (23)

What is the message this arboreal sentinel delivers to the sole surviving member of the Hoel clan via branches that "wave [like] semaphores" against the winter sky? It seems to be one of strange comfort: the thought that one of the most terrible moments in Nick's existence will be experienced by the tree dispassionately as just another "insignificant," "transitory" micro-moment in the yawning expanse of its life, one that will become inscribed as another ring in a sublime succession of rings. Similarly, when his plane is shot down during the Vietnam war, a seriously wounded, parachuting Doug Pavlicek is saved by a massive banyan fig that has grown over the course of "centuries" wherein human conflicts and suffering comes and goes. In the fig's lifetime, it has witnessed "[w]ar on the backs of elephants [that] gave way to televised moon landings and hydrogen bombs" (81), yet through all this it endures and is worshipped by Thai pilgrims as a holy site. After Neelay Mehta falls from an oak tree "with phantasmagoric branches growing since California was Spain's remotest outpost" (101), the newly paralyzed boy regards this entity with dispassionate awe as "a colossal, rising, reaching, stretching space elevator of a billion independent parts … the most perfect piece of self-writing code that his eyes could hope to see" (103). And, as we have seen, Patricia Westerbrook's *The Secret Garden* addresses our long-term evolutionary kinship with trees years after her traumatic encounter with the destruction of an aspen grove, a process whereby "a forest, spreading across these mountains since before humans left Africa [was] giving way to second homes" (133).

Two of the most significant examples of arboreal sublimity as a response to trauma occur near the end of *The Overstory*. In moments of acute anxiety stemming from their former eco-activism, both Adam Appich and Mimi Ma experience strange visions in which the Earth's deep past becomes indistinguishable from its state in the deep future. Adam's vision occurs while he is under house arrest in his Manhattan apartment, awaiting sentencing for acts of "domestic terrorism" (470). Having just had a terrible fight with his wife Lois regarding whether his decision not to inform on Mimi and Nick is justified given the price his family will pay if he's convicted, Adam falls into a fitful sleep, only to be awakened by pain in his lower back and bladder. As his stumbles across his apartment, he catches a glimpse of "a pack of gray wolves com[ing] down Waverly from Washington Square, chasing a white-tailed deer" (462) before he injures himself further by falling and hitting his mouth on a windowsill. As he rises from the floor "stupid with agony" he looks out the window and has an otherworldly vision:

> The moon shines over the tip of a tree-covered island. Brick, steel, and right angles give way to moonlit, mounded green. A stream runs through a ravine that cuts towards West Houston. The towers of the Financial District are gone, changed into wooded hills … a forest as dense, terrifying and inescapable as childhood. Arboretum America … hornbeam, oak, cherry, half a dozen kinds of maple. Honey locusts armored with thorns against extinct megafauna… Wilderness rushes down lower Broadway, the island as it was a thousand years back or a thousand years on.
>
> (462–3)

Significantly, in Adam's vision of Manhattan rewilded, no human residents are present; instead, wolves, deer, owls, bears and sea turtles populate this old/new "wilderness." And trees once again reign supreme, including species such as "pignut hickories" that are (in the twenty-first century) relatively uncommon. The question arises as to why, in this moment of extreme "stress [over] his arrest," Adam should have this vision of "Arboretum America." If this is a vision of the future, how could this be the case, given ways it also resembles America before European colonization? Moreover, why doesn't this vision of America 1,000 years ago include the presence of Native Americans, who are thought to have inhabited North America as far back as 16,000 to 20,000 years ago? One way of accounting for this is that Adam's vision is consistent with a romantic, Eurocentric one: sublime wilderness as it was venerated by late eighteenth- and early nineteenth-century white settler colonists as a place supposedly devoid of human presence or culture. The ideological erasure of indigenous peoples in Adam's vision is consistent with romantic constructions of wilderness critiqued by William Cronon in "The Trouble with Wilderness." Likewise, the only way his vision could be New York in the distant future is if some catastrophe occurred, wherein the city has been erased and replaced by a new "empire" of trees and wildlife. This would be consistent with Adam's belief that one day his activism will be justified by a future "cataclysm" that he is sure will come "long before his seventy plus seventy years [prison sentence] is up" (481). Thus, an apocalyptic "wiping clean" of the slate is married with a deep ecological vision of wilderness renewed in the absence of mass humanity. Meanwhile, in the present day at his sentencing hearing, Adam psychologically escapes from his fate with thoughts of sublime, arboreal time:

> *Seventy plus seventy is nothing. A black willow plus a wild cherry.* He was thinking oak. He was thinking Douglas-fir or yew. Seventy plus seventy. With reductions for good behavior, he might even finish out the first half of the sentence just in time to die.
>
> (471, emphasis in original)

A similar vision of the future/past occurs when Mimi Ma "gets enlightened" (499) while leaning against a pine tree in San Francisco's Mission Dolores Park after hearing the news about Adam's sentencing and deducing that Doug has "sold two lives for hers" (490). Gripped by guilt over the fact that her freedom has been purchased by Doug and Adam's incarceration, she considers turning herself in to take responsibility for her role in the group's monkey-wrenching and in Olivia's death. However, she ends up deciding against this course of action. Rejecting the "frantic need to steer" (499) her life, Mimi embraces the wisdom of the Arhats depicted on her family's ancient scroll and listens to the voice of trees that tell her "*[a] thing can travel everywhere, just by holding still*" and that she should not reject suffering but should be content just to "*[b]e still and feel*", (500 emphasis in original). It is at this moment of intense pain and self-conflict that she has a vision that resonates with Adam's:

> The fires will come, despite all efforts, the blight and windthrow and floods. Then the Earth will become another thing, and people will learn it all over again. The vaults of seed banks will be thrown open. Second growth will rush back in, supple, loud, and testing all possibilities. Webs of forest will swell with species shot through in shadow and dappled by new design. Each streak of color on the carpeted Earth will rebuild its pollinators. Fish will surge again up all the watersheds, stacking themselves as thick as cordwood through the rivers, thousands per mile. Once *the real world* ends.
>
> (500, emphasis in original)

As in Adam's vision, the pre-requisite for a renewed Earth in the distant future is catastrophe in the near term: a period of apocalyptic "fires ... blight ... windthrow and floods" despite all efforts (including those of eco-activists) to avert this destruction. Presumably, many of the Earth's entangled lifeforms will suffer greatly before a new world can grow up from the ruins of the old. But, significantly, "people" are imagined to be part of this future; a term which (in our time) typically applies to human beings but may, in the future, encompass other life forms. The inclusion of humans (and other beings) who will re-learn the Earth "all over again" in the future distinguishes Mimi's vision in important ways from Adam's. Indeed, human beings with the foresight to create seed banks help to re-populate this new Earth with plants, including trees which "swell" in entangled forests. From a human perspective, re-education is needed in order to accomplish this renewal, a capacity to imagine the "real world" we currently inhabit—a capitalist one that denies entanglement is a key to survival—*is not the best of all possible worlds.* Rather than the return to colonialist "wilderness" imagined in Adam's vision, Mimi's suggests a more hopeful future could be built by all peoples working in harmony within life's larger "webs" of life.

So, which vision of the future/past—Adam's or Mimi's—will take root? As with the novel's other fractal narratives both outcomes are equally possible, held in suspension by the novel's "conclusion." Read optimistically, such suspended possibilities suggest an *open* future, a timeline which is not (always) already determined by economics and which therefore constitutes a challenge to *build* a new "real world" that can support our species and others. Interpreted less generously, one could read this suspension as a form of equivocation given that the novel doesn't provide a coherent vision of how global capitalism could be radically reformed or replaced by a more sustainable economic system, no sense of how a new world could be achieved short of systemic collapse and ecological catastrophe as pre-requisites for renewal. Although characters like the intellectual property lawyer Ray Brinkman contemplate possible legal reforms such as the extension of personhood to non-human entities (including trees on p. 250), there is no character in the novel who voices the perspective of capitalist hegemony or through whom a systematic critique of the economic juggernaut might be explored. Instead of hazarding such things, the text offers the familiar terrain of deep-ecological Romanticism, including a conviction that regardless of what kind of world emerges in the future, trees will be a prominent presence (as they are in both Adam and Mimi's visions). The novel suggests that the crisis we must negotiate is not whether trees will endure, but whether human beings will survive into the future. Hence, Nick's belated revelation about what Olivia's "voices must always have meant": "*The most wondrous products of four billion years of life need help.* Not them; us. Help from all quarters" (493).

Thus, the voice of deep time is evoked to minimize the scope of damage Anthropocene-era capitalism can inflict on nature *in the long term*, and comfort in the thought that arboreal life will "*still*" endure as embodiments of "the word life has been saying, since the beginning" and that "*will never end*" (502, emphasis in original). But if entanglement teaches us anything it is that our fate as a species cannot be so readily extricated from that of other life on the planet. Depending on the severity of the damage "we" (consumers in developed nations) inflict today, life as we know it on planet Earth could take a very, very long time to rebound, many species could be lost never to return and vast numbers of our human kin could suffer or perish as well.

Epilogue: Does Entanglement Constitute a Fertile Ground for Future Ecopolitics?

Studied together, viral capitalism and arboreal entanglement suggest both the benefits and drawbacks of a flat ontology as a basis for formulating new environmental politics in the Anthropocene era. On the one hand, literary depictions of our material entanglement with other species illustrates what the editors of *Arts of Living on a Damaged Planet* characterize as our "monstrous" condition of needing to come to terms with "entanglement as life and as danger" (Tsing et al. M4). Specifically, in this chapter I have explored ways that entanglement is both an evolutionary fact of "life" and (simultaneously) a source of "danger" in viral capitalist praxis that discounts such enmeshment, creating increased risk in the process. On the other hand, flat ontologies that do not explore (or even deny) what is distinctive about human agency potentially circumvent the issue of our species' unique *responsibilities* vis-à-vis the more-than-human world. Because human beings have the capacity to understand deep time (through complex communication systems such as science and art) and the long-lasting effects of anthropogenic changes we are making to the Earth's ecosystems, we require a politics that not only does justice to our ontological entanglements but also recognizes why our social systems have been *resistant* to accepting materialist ecology's insights, along with exploring possible solutions to such problems. At issue here is a version of the "is"/ "ought" dichotomy that emerges in modern, secular societies: in the case of environmental politics, the gap between what science tells "is" factually true about our ecological entanglements versus the social negotiation of what our societies "ought" to do based upon such insights (within the domain of ethics and politics). Hence, Hannes Bergthaller and Eva Horn express skepticism about the utility of "flat ontology" (10–11) as a basis for effective politics in *The Anthropocene: Key Issues for the Humanities.*

In "Against Hybridism," Andreas Malm goes so far as to suggest that the kind of "hybridism" evident in new materialist philosophies (exemplified in the work of

Latour, Moore and Val Plumwood) which "rush[es] to jettison the categories [of nature and society] as soon as the extent of their entanglement comes into view—is, at a closer look, merely the flipside of substance dualism" (162): i.e. is a form of "substance" and "property" *monism* (164) that refuses to acknowledge any distinction between human and more-than-human ontologies but, instead, sees these modes of being as an "undifferentiated oneness." As an alternative to this, Malm advocates for a viewpoint that embraces "substance" monism (i.e. acknowledges that nature and society are materially entangled) but also recognizes distinctive, emergent "properties" that characterize humanity's "social" systems, in the same way that what we call "mind" is at once something irreducibly dependent upon embodiment, but which also cannot be fully reduced to embodiment insofar as it is "an outgrowth of that substance in possession of unique mental properties" (166). Malm claims that historical materialism represents exactly this kind of material monist–property dualist perspective, illustrating this principle with a trope relevant to our discussion of *The Overstory*:

> Picture a tree. It grows out of the soil, draws nourishment from it, expires the moment it is cut off from it: yet it cannot be reduced to it. Nature is a soil for society, the fold out of which it grew and the envelope it can never break out of, but just as a tree can be distinguished from its soil, society can be differentiated from nature because it has shot up from the ground and branched off in untold directions over the course of what we refer to as history.
>
> (166–7)

Malm argues persuasively that there is a need to distinguish ways human "society can be differentiated from nature" in order to analyze phenomena such as the anthropogenic sources of climate disruption and fossil capitalism's *responsibility* as a "perturbation" (171) that triggers cascading climate effects. In an undifferentiated monistic account of climate change, it is difficult to see how one could isolate agency (and require accountability) for specific forces driving change in such a complex phenomenon or pinpoint specific areas that require reform.

On the other hand, Malm's critique of hybridism doesn't acknowledge ways that new materialist thinkers do not necessarily erase *all* distinctions between human and more-than-human agents in their accounts of entanglement. The absence of any serious engagement with Donna Haraway's work in Malm's essay is odd, considering that "Against Hybridism" was published in 2019, three years after Haraway's *Staying with the Trouble*. In this text Haraway is explicitly concerned with *storytelling* as an especially impactful mode of sympoiesis. Although she claims that Thom van Dooren's *Flight Ways* demonstrates that "storying [as multimodal semiotics] cannot any longer be put into the box of human exceptionalism" (35), she also acknowledges the importance of *specificity* (focus, detail, place-based awareness) and *care* (response-ability) in the kinds of storying we humans engage in, implicitly acknowledging important differences in the positionality and purpose of different kinds of discourse. This is why Haraway repeatedly asserts that "it matters what thoughts think thoughts. It matters what knowledges know knowledges. It matters what relations relate relations. It matters what words world worlds. It matters what stories tell stories" (35). Instead of reflecting an undifferentiated monism which makes no distinction between different kinds of "storying," *Staying with the Trouble* can be seen as a passionate call to "change the story" (40) or, more accurately the *stories* about nature and culture that those of us living under global capitalism have inherited, particularly subjects of the global north that have benefitted the most from its patriarchal, colonialist and speciesist narratives.

Literature is one place where we can collectively revise these stories, re-thinking the grounds for ongoingness. *The Overstory*'s concluding paragraph hints at this by trying to mobilize paradoxes in the word "still." The artist Nick Hoel (with the help of some unnamed Native Americans) has spelled out this word on a gigantic scale, using downed wood on a hillside in an unnamed "northern" forest. As a signifier, "still" can mean static, unmoving; or the kind of transcendental "stillness" Mimi Ma achieves via enlightenment, wherein "a thing can travel everywhere, just by holding still" (500). Yet neither of these meanings fully reflect what is happening to Hoel's piece of long-term performance art, because materially nature is *never still*: "[a]lready this word is greening" by nourishing lichen, fungi and insects such that in "[t]wo centuries … these five living letters, too, will fade back into the swirling patterns, the changing rain and air and light" (502). Is the story told here "still" the familiar one of ephemeral human agency in contrast to nature's beautiful and permanent forms that "will never end"? Or, more darkly, "still" a colonialist one wherein a white savior's techne is accomplished only by harnessing the labor of "cheap natures" (anonymous native peoples and forests)? The proximity of "methane-belching tundra" near Hoel's word sculpture could be read as qualifying Romantic deep time's transcendence of the human by reminding us that global warming will "still" linger long into the future even if our individual agency is ephemeral. We might "still" be able to avert long-term catastrophe if we can excavate colonialism's lingering legacy in the stories we tell (even those as accomplished as Powers's) and reform the history of multispecies injustice in order to imagine different narratives about the future. A new story about our species' relationship to time? If this seems like a tall order, it is "still" better (in my view) than hoping that *deus ex machina* technology will save us, whether it be geoengineering or Powers's AI "learners" that dispassionately observe Hoel's artwork from a perspective far above the Earth.

Notes

1 See Gabriel Popkin's "Wood Wide Web—The Underground Network of Microbes That Connects Trees—Mapped for the First Time" in *Science* magazine.

2 For a representative text, see Bookchin's *Re-enchanting Humanity: A Defense of the Human Spirit against Anti-Humanism, Misanthropy, Mysticism and Primitivism*, particularly Chapter 2, "From 'Selfish Genes' to 'Mother Gaia.'" Bookchin has been a vocal critic of scientists and post-humanist environmentalists who, in his view, falsely and reductively suggest analogies between human culture and more-than-human phenomena.

3 For example, critics such as Greg Garrard and Hannes Bergthaller develop interpretations that suggest Atwood believes in a trans-historical "human nature" that's the product of biological evolution (Garrard) and/or that requires "self-domestication"

(Bergthaller). For more on this, see Garrard's "Reading as an Animal" and Bergthaller's "Housebreaking the Human Animal." However, other scholars such as Gerry Canavan read the trilogy's first two novels as satirizing biological essentialism, such that "Crake's totalizing genetic determinism" in his scheme to replace our species with the Crakers "is just another part of the joke" ("Hope" 152). As is often the case with creative writers, Atwood's own published comments and interviews do not help to clarify her stance on this issue. For example, in "Dire Cartographies" she suggests that the Crakers represent "a little attempt at [creating] utopia" within a broader dystopia, insofar as they are "a group of quasi-humans who have been genetically engineered so that they will never suffer from the ills that plague *Homo sapiens*" (91) and goes on to refer to humans that survive Crake's pandemic as potentially being "a dystopic threat to the tiny utopia of genetically modified, peaceful and sexually harmonious New Humans that is set to replace them" (93). Such descriptions suggest that Atwood would locate the source of human ills in our genetic makeup and stresses bioengineering's "utopian" potential to correct these problems (a physiological solution that negates the necessity of freely chosen reforms). However, in interviews such as one for the CBC that Atwood gave while participating in a 2013 "Thinking Extinction" conference, she stressed the critical importance of our collective free will in shaping future biodiversity, rejecting the thesis that such diversity is "doomed" in favor of a perspective that the ethical choices we make in the present "can have a real effect" on the future.

4 For my analyses of the Romantic sublime, see "'Mont Blanc' and the Sublimity of Materiality" and *The Ecology of Wonder in Romantic and Postmodern Literature.*

References

Adams, Rachel. An Overstory for Our Time. *American Literature*, Vol. 92, No. 4, December 2020, pp. 799–807.

Alaimo, Stacy. *Bodily Natures: Science, Environment, and the Material Self.* Indiana University Press, 2010.

Appleton, Sarah. Corp(Se)ocracy: Marketing Death in Margaret Atwood's Oryx and Crake. *Latch*, Vol. 4, 2011, pp. 63–73.

Atwood, Margaret. Dire Cartographies: The Roads to Ustopia. In *Other Worlds: SF and the Imagination.* Anchor Books, 2012, pp. 66–96.

Atwood, Margaret. *Oryx and Crake.* Anchor Books, 2004.

Atwood, Margaret. *MaddAddam.* Nan A. Talese, 2013.

Atwood, Margaret. *The Year of the Flood.* Anchor Books, 2010.

Bacigalupi, Paolo. *The Windup Girl.* Night Shade Books, 2009.

Bergthaller, Hannes. Housebreaking the Human Animal: Humanism and the Problem of Sustainability in Margaret Atwood's Oryx and Crake and The Year of the Flood. *English Studies*, Vol. 91, No. 7, November 2010, pp. 728–43.

Bergthaller, Hannes and Eva Horn. *The Anthropocene: Key Issues for the Humanities.* Routledge, 2020.

Bookchin, Murray. *Re-enchanting Humanity: A Defense of the Human Spirit against Anti-Humanism, Misanthropy, Mysticism and Primitivism.* Cassell, 1995.

Bradley, James. *Clade.* Titan Books, 2017.

Canavan, Gerry. Hope, But Not for Us: Ecological Science Fiction and the End of the World in Margaret Atwood's Oryx and Crake and The Year of the Flood. *Literature Interpretation Theory*, Vol. 23, 2012, pp. 138–59.

Cole, Matthew Benjamin. "At the Heart of Human Politics": Agency and Responsibility in the Contemporary Climate Novel. *Environmental Politics*, Vol. 31, No. 1, 2022, pp. 132–51.

Conrad, Joseph. *Heart of Darkness* (Norton Critical Edition). 5th ed., edited by Paul B. Armstrong. Norton, 2017.

Cronon, William. The Trouble with Wilderness. In *Uncommon Ground: Toward Reinventing Nature.* Norton, 1995, pp. 69–90.

DeLuca, Laura. Temporality in *The Overstory* by Richard Powers. *The Explicator*, Vol. 80, No. 3, pp. 78–80.

Economides, Louise. *The Ecology of Wonder in Romantic and Postmodern Literature.* Palgrave Macmillan, 2016.

Economides, Louise. "Mont Blanc" and the Sublimity of Materiality. *Cultural Critique*, Vol. 61, Fall 2005. pp. 87–114.

Garrard, Greg. Reading as an Animal: Ecocriticism and Darwinism in Margaret Atwood and Ian McEwan. *Local Natures, Global Responsibilities: Ecocritical Perspectives on the New English Literatures.* Brill, 2010, pp. 223–42.

Hageman, Andrew. The Challenge of Imagining Ecological Futures: Paolo Bacigalupi's *The Windup Girl. Science Fiction Studies*, Vol. 39, No. 2, July 2012, pp. 283–303.

Haraway, Donna. *Staying with the Trouble: Making Kin in the Chthulucene.* Duke University Press, 2016.

Heise, Ursula K. What's the Matter with Dystopia? *Public Books*, February 1, 2015, www.publicbooks.org/whats-the-matter-with-dystopia/.

Heller, Peter. *The Dog Stars.* Vintage Books, 2012.

Hitt, Christopher. Toward an Ecological Sublime. *New Literary History*, Vol. 30, No. 3, 1999, pp. 603–23.

Latour, Bruno. Is This a Dress Rehearsal? *Critical Inquiry*, Vol. 47, No. S2, Winter 2021, pp. S25–27.

Lyotard, Jean-François. *The Inhuman: Reflections on Time.* Translated by Geoffrey Bennington and Rachel Bowlby. Stanford University Press, 1988.

Malm, Andreas. Against Hybridism: Why We Need to Distinguish Between Nature and Society, Now More than Ever. *Historical Materialism*, Vol. 27, No. 2, 2019, pp. 156–87.

Malm, Andreas. *Corona, Climate, Chronic Emergency: War Communism in the Twenty-First Century.* Verso, 2020.

Martin, Javier. Dystopia, Feminism and Phallogocentrism in Margaret Atwood's Oryx and Crake. *Open Cultural Studies*, Vol. 3, No. 1, 2019, pp. 174–81, www.degruyter.com/document/doi/10.1515/culture-2019-0015/html?lang=en.

McCarthy, Cormac. *The Road.* Vintage Books, 2006.

Miller, Elizabeth Carolyn. Dendography and Ecological Realism. *Victorian Studies*, Vol. 58, No. 4, Summer 2016, pp. 696–718.

Moore, Jason W. *Capitalism in the Web of Life: Ecology and the Accumulation of Capital*, Verso, 2015.

Popkin, Gabriel. Wood Wide Web—The Underground Network of Microbes That Connects Trees—Mapped for the First Time. *Science*, May 15, 2019, www.science.org/content/article/wood-wide-web-underground-network-microbes-connects-trees-mapped-first-time.

Powers, Richard. *The Overstory.* Penguin Random House, 2018.

Quammen, David. *The Tangled Tree: A Radical New History of Life.* Simon & Schuster, 2018.

Schoene, Berthold. Arborealism, or Do Novels Do Trees? *Textual Practice*, Vol. 36, No. 9, 2022, pp. 1435–58.

Selisker, Scott. "Shutter-Stop Flash-Bulb Strange": GMOs and the Aesthetics of Scale in Paolo Bacigalupi's *The Windup Girl. Science Fiction Studies*, Vol. 42, No. 3, November 2015, pp. 500–18.

Shiva, Vandana. *Biopiracy: The Plunder of Nature and Knowledge.* South End Press, 1997.

Shiva, Vandana. Vandana Shiva on Sexual Economics, Biopiracy and Women's Ongoing Resistance to Colonialism. Interviewed by Nandita Sharma and Allison Campbell, *Atlantis*, Vol. 23, No. 2, Spring/Summer/Spring/Summer 1999, pp. 67–74.

Sullivan, Heather I. Cross-Infections of Vegetal–Human Bodies in Science Fiction. *Science Fiction Studies*, Vol. 49, No. 2, July 2022, pp. 342–58.

Tsing, Anna. *The Mushroom at the End of the World: On the Possibility of Life in Capitalist Ruins*. Princeton University Press, 2015.

Tsing, Anna et al. *Arts of Living on a Damaged Planet*. University of Minnesota Press, 2017.

Wallace-Wells, David. *The Uninhabitable Earth: Life after Warming*. Tim Duggan, 2019.

Žižek, Slavoj. *Pandemic! Covid-19 Shakes the World*. OR Books, 2020.

Index

Note: Page numbers with an "n" and note number after the page reference refer to notes.

FSC
www.fsc.org
MIX
Papier | Fördert gute Waldnutzung
FSC® C083411

Kolibri 360 GmbH
Ferdinand-Jühlke-Straße 7
99095 Erfurt, Deutschland
produktsicherheit@kolibri360.de